# Machine Learning for Auditors

## Automating Fraud Investigations Through Artificial Intelligence

Maris Sekar

Apress®

***Machine Learning for Auditors: Automating Fraud Investigations Through Artificial Intelligence***

Maris Sekar
Calgary, AB, Canada

ISBN-13 (pbk): 978-1-4842-8050-8 ISBN-13 (electronic): 978-1-4842-8051-5
https://doi.org/10.1007/978-1-4842-8051-5

Copyright © 2022 by Maris Sekar

This work is subject to copyright. All rights are reserved by the Publisher, whether the whole or part of the material is concerned, specifically the rights of translation, reprinting, reuse of illustrations, recitation, broadcasting, reproduction on microfilms or in any other physical way, and transmission or information storage and retrieval, electronic adaptation, computer software, or by similar or dissimilar methodology now known or hereafter developed.

Trademarked names, logos, and images may appear in this book. Rather than use a trademark symbol with every occurrence of a trademarked name, logo, or image we use the names, logos, and images only in an editorial fashion and to the benefit of the trademark owner, with no intention of infringement of the trademark.

The use in this publication of trade names, trademarks, service marks, and similar terms, even if they are not identified as such, is not to be taken as an expression of opinion as to whether or not they are subject to proprietary rights.

While the advice and information in this book are believed to be true and accurate at the date of publication, neither the authors nor the editors nor the publisher can accept any legal responsibility for any errors or omissions that may be made. The publisher makes no warranty, express or implied, with respect to the material contained herein.

Managing Director, Apress Media LLC: Welmoed Spahr
Acquisitions Editor: Jonathan Gennick
Development Editor: Laura Berendson
Coordinating Editor: Jill Balzano

Cover image designed by Freepik (www.freepik.com)

Distributed to the book trade worldwide by Springer Science+Business Media LLC, 1 New York Plaza, Suite 4600, New York, NY 10004. Phone 1-800-SPRINGER, fax (201) 348-4505, e-mail orders-ny@springer-sbm.com, or visit www.springeronline.com. Apress Media, LLC is a California LLC and the sole member (owner) is Springer Science + Business Media Finance Inc (SSBM Finance Inc). SSBM Finance Inc is a **Delaware** corporation.

For information on translations, please e-mail booktranslations@springernature.com; for reprint, paperback, or audio rights, please e-mail bookpermissions@springernature.com.

Apress titles may be purchased in bulk for academic, corporate, or promotional use. eBook versions and licenses are also available for most titles. For more information, reference our Print and eBook Bulk Sales web page at http://www.apress.com/bulk-sales.

Any source code or other supplementary material referenced by the author in this book is available to readers on GitHub via the book's product page at https://github.com/Apress/machine-learning-for-auditors.

Printed on acid-free paper

*To my lovely wife Anusha*
*and my children Adi and Arya*
*for giving me valuable family time*
*to pursue my dreams.*

# Table of Contents

# About the Author

**Maris Sekar** is a professional computer engineer, Senior Data Scientist (Data Science Council of America), and Certified Information Systems Auditor (ISACA). He has a passion for using storytelling to communicate on high-risk items within an organization to enable better decision-making and drive operational efficiencies. He has cross-functional work experience in various domains such as risk management, oil and gas, and utilities. Maris has led many initiatives for organizations such as PricewaterhouseCoopers LLP, Shell Canada Ltd., and TC Energy. His love for data has motivated him to win awards, write articles, and publish papers on applied machine learning and data science.

# About the Author

# About the Technical Reviewer

**Dr. David Paper** is a retired academic from the Utah State University's (USU) Data Analytics and Management Information Systems Department in the Huntsman School of Business. He has over 30 years of higher education teaching experience. At USU, he taught for 27 years in the classroom and distance education over satellite. He taught a variety of classes at the undergraduate, graduate, and doctorate levels, but he specializes in applied technology education.

Dr. Paper has competency in several programming languages, but his focus is currently on deep learning with Python in the TensorFlow-Colab Ecosystem. He has published extensively on machine learning, including such books as *Data Science Fundamentals for Python and MongoDB* (2018, Apress), *Hands-on Scikit-Learn for Machine Learning Applications: Data Science Fundamentals with Python* (2019, Apress), and *TensorFlow 2.x in the Colaboratory Cloud: An Introduction to Deep Learning on Google's Cloud Service* (2021, Apress). He has also published more than 100 academic articles.

Besides growing up in family businesses, Dr. Paper has worked for Texas Instruments, DLS Inc., and the Phoenix Small Business Administration. He has performed Information Services (IS) consulting work for IBM, AT&T, Octel, the Utah Department of Transportation, and the Space Dynamics Laboratory. He has worked on research projects with several corporations, including Caterpillar, Fannie Mae, Comdisco, IBM, RayChem, Ralston Purina, and Monsanto. He maintains contacts in corporations such as Google, Micron, Oracle, and Goldman Sachs.

# About the Technical Reviewer

# Introduction

"Have you ever pondered over missing a fraudulent transaction? What if there was a smart person out there who found a way to hide a single financial transaction among millions of other transactions? Can you really find a needle in a haystack?" These thoughts echoed within me while I was trying to come up with an audit plan to identify transactions that looked suspicious.

Internal auditors are the last line of defense when it comes to protecting an organization's confidentiality, integrity, and availability of information assets. Auditors are often portrayed as professionals verifying if processes, procedures, and controls are in place to mitigate the business risks. What if auditors can also bring in additional value to the organization as a trusted advisor?

The complexity and size of organizational information assets are ever increasing. The chance of finding fraudulent transactions using traditional audit methods is small and decreases over time. Scalable and sustainable solutions need to be incorporated by the internal audit department in order to make a significant advance in the fight against the organizational threat actors.

With the help of artificial intelligence, machine learning, and data science, the challenges faced by traditional audit methods can be overcome. This book provides a systematic way to apply artificial intelligence to the field of internal audit. It is especially useful for those with domain expertise in the field of auditing, but lack the technical expertise required for artificial intelligence.

The book is structured to introduce the reader to internal auditors' evolved role as trusted advisors. The current challenges and solutions faced by the internal audit team are illustrated. This is followed by a primer to machine learning and data science concepts. Data visualization and storytelling are then covered in a great level of detail to support the sharing of audit results. Finally, practical applications of the artificial intelligence concepts are described and demonstrated in terms of recipes. Each recipe is accompanied with working code examples that can be easily incorporated into audits.

I hope you have as much fun reading it and applying it in your practice, as I had in writing the book.

# PART I

# Trusted Advisors

CHAPTER 1

# Three Lines of Defense

*Artificial intelligence* (AI) and *machine learning* (ML) have come far from mere buzzwords to delivering tangible business value. Many organizations have successfully leveraged AI and ML in multiple practical use cases. Risk management poses a unique opportunity for AI applications. In addition to helping organizations manage risks proactively, internal auditors can use AI to align and realize business objectives. Governance and role definition have become the central topic of concern in risk-management applications due to auditors' need to maintain their independence. Particularly, auditors should not audit their own work. Banks have been at the forefront of these risk management applications (to detect and deter fraud).

The "*Three Lines of Defense*" model has been widely adopted to help guide AI implementations in the field of risk management, including auditing. This model is a framework that clarifies the roles and responsibilities for effective risk management and control. In 2020, amid growing corporate risks worldwide, the "Three Lines of Defense" model was evolved and refined as the "*Three Lines*" model. This chapter will look at why governance and clearly defined roles are essential for utilizing AI and ML in risk management. It also makes a solid case for why the current time, with the "Three Lines" model, empowers the use of ML in auditing.

## AI, ML, and Auditing

Artificial intelligence and machine learning are used interchangeably. It is essential to clarify what they mean. Imagine a child playing with Mega Blocks. The child makes decisions on what block goes on top of another block. The decisions are driven through past experiences (or learning). For instance, the child knows square blocks cannot go on top of the triangle blocks after trying it several times. Here, the child's decision-making mechanism to decide which block goes next can be modeled by machine learning. If the child is replaced with a robot, then the robot as a whole becomes what is known as an artificial intelligence agent. **AI** consists of programming a machine to do tasks that humans typically carry out.



© Maris Sekar 2022
M. Sekar, *Machine Learning for Auditors*, https://doi.org/10.1007/978-1-4842-8051-5_1

Machine learning is divided into two main categories - *supervised learning* and *unsupervised learning*. In supervised learning, the labeled inputs and expected output of the process being modeled are used to train the model. In contrast, unsupervised learning involves the use of self-learning techniques for modeling an unlabeled dataset. In our example with the child and the Mega Blocks, supervised learning is when the child learns from their parent, showing them how to use the blocks to build a tower. Unsupervised learning is when the child learns to build a house that has never been constructed before based on their own understanding by experimenting with different constructs. The difference between supervised and unsupervised learning is an important distinction, and it will be used to explain some of the challenges of supervised learning.

The term "auditor" means different things in different situations, but we will refer to it to represent the "internal auditor" in this book. A company employs an internal auditor to inspect the company's processes and understand the organization's risks. The auditors then gather the risks through risk assessments (often conducted on an annual basis) with senior management and the board of directors. As part of the verification, auditors assess the high-risk company processes, and the supporting documentation (including organizational data) is examined.

AI and ML have posed unique challenges due to the internal auditor's need to maintain independence. According to the Institute of Internal Auditors (IIA) Guidance, "Independence is the freedom from conditions that threaten the ability of the internal audit activity to carry out internal audit responsibilities in an unbiased manner." The developmental effort associated with an AI or ML application might be seen as a barrier to internal audit if internal audit developed the solution themselves. Other roles need to be defined to support the development process to maintain the internal audit's oversight and independence. A proper governance structure must also be put in place to support collaboration and information sharing among the AI/ML developers and the internal audit function.

Another challenge for the internal audit function is the quality of data used in their AI/ML models. Some applications of machine learning, like supervised machine learning, make decisions based on historical data. There is a risk that internal audit may decide based on "dirty" and biased data. Although these problems are not new, it may be a more significant issue with AI/ML modeling. In collaboration with data analysts, the auditors can check the data integrity before the application processes the data. Data bias is a more complicated problem to overcome. During the data exploration step in the AI process, data specialists can "clean" the data and monitor outliers by comparing and validating with the rest of the population. Besides, the roles and responsibilities for internal auditors and the separate AI governance entity can be clearly defined to ensure the internal audit's independence is maintained.

In the next section, we will look at the "Three Lines of Defense" model. The model will be explored in detail, including where it falls short in risk management applications.

## The Three Lines of Defense Model

The "Three Lines of Defense" model was introduced in 2013 by IIA to clarify the roles and responsibilities to manage and control risks effectively. The roles themselves have existed before 2013, but they have not been formally defined and discussed in the level of detail as is described here.

Figure 1-1 shows the Three Lines of Defense model stated by the Institute of Internal Auditors in a position paper titled "The Three Lines of Defense in Effective Risk Management and Control" in January of 2013. It was adapted from the "Three Lines of Defence" model from the European Confederation of Institutes of Internal Auditing (ECIIA)/Federation of European Risk Management Associations (FERMA) *Guidance on the 8th EU Company Law Directive, article 41*.

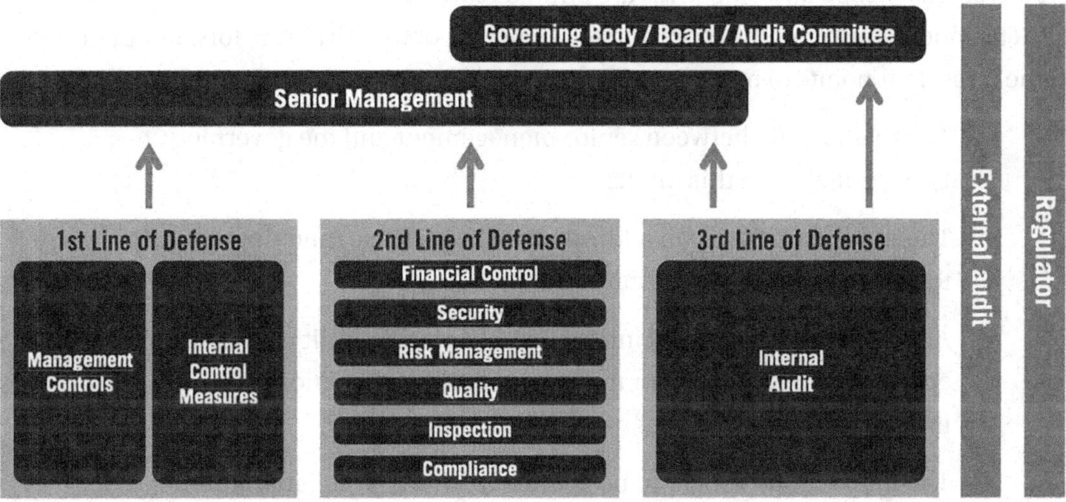


***Figure 1-1.*** *The Three Lines of Defense model*

The three lines of defense illustrated in Figure 1-1 are defined as follows:

- **1st Line of Defense.** This represents operational management's role in assessing, controlling, and mitigating risks in the form of management and internal controls. The operational management team's report to senior management.
- **2nd Line of Defense.** It consists of financial control, security (physical and information), risk management, quality, inspection, and compliance functions, along with their roles in ensuring the risks within their functional domain are managed and mitigated effectively. These functions report to senior management as well.
- **3rd Line of Defense.** The internal auditors act as the 3rd line of defense in that they independently assess the organization's risks, both internal and external, and ensure the risks are managed effectively. The internal audit team works with senior management to mitigate the risks and report to the audit committee/governing body/board of directors.

There is also an external audit function recognized through this model that can act as a fourth line of defense. The external auditors provide another independent assurance. However, their scope and objectives are limited to the financial reporting risks.

The model illustrates the reporting relationships of the various actors, but there are some important points to note:

- The relationship between senior management and the governing body is unclear in this model.
- The model is focused on addressing risks (defense) and does not touch on meeting organizational objectives.
- Although there is upward interaction by internal audit with the governing body and senior management, the interaction between internal audit and the other two lines is not present.
- The interactions appear to be in one direction. To collaborate effectively, information sharing/interaction needs to take place in both directions. For example, during the risk assessment process, the governing body provides inputs for enterprise-wide risks to internal

audit. Internal audit assesses the risks and provides a report on the findings to management and the audit committee. This shows the interaction in both directions.

In the next section, we will explore the complexities and opportunities of risk management in general.

## Risk Management Complexities

Organizations use a complex array of people, processes, and technology in order to manage and mitigate the risks introduced by their internal and external processes. There are process owners or service providers who are accountable for their processes and assets. Process owners manage risks associated with their operations. Process owners and product/service providers within an organization are the first line of control ("*first line of defense*") to know what risks affect their products and services. Personnel belonging to functions such as security (physical and cyber), compliance, quality, financial control, and risk management provide monitoring, expertise, and support based on their domain knowledge. These domain experts help in overseeing the process owners and ensuring the risks are controlled as part of their roles. Hence, they are known as the "*second line of defense.*" Finally, internal auditors serve as an independent assurance provider function. They ensure there is oversight over the domain experts when it comes to identifying and managing risks. This is the reason for calling them the "*third line of defense.*" Process owners and domain personnel report to senior management, who in turn report to the governing body. The internal auditors, on the other hand, report directly to the governing body or the board of directors and exercise their independence. An important point to note is that independence does not mean isolation from the other two lines of defense. A great amount of coordination is required between these entities in order for them to be effective in achieving the overall organizational goals.

Another complexity associated with these "lines of defense" is the level of access each of them has. Process owners will have full administrative privileges to a particular process and can read and write data. Financial control personnel will only have read access to the financial data of the same process. Internal auditors will often have full visibility of the entire system but may lack the process or domain knowledge that is needed to better understand the system. For example, for a materials management process, the process owner will have full access (read/write) to the data for the materials management process. The cybersecurity team will have access (read-only) to the security component of the Enterprise Resource Planning (ERP) system the process resides in. The internal audit function within a company will have unfettered access (read-only) to the ERP and the material management data. Having the domain knowledge and access to data for both the process and the ERP system it resides in can be an invaluable tool in providing assurance to the company's stakeholders as well as aid in the achievement of business objectives. This will be discussed in detail later in this chapter in terms of offense and defense opportunities that are available for internal auditors to choose from.

An organization may have hundreds if not thousands of IT applications. Each application has its own application-generated data (application logs) and system-generated data (system logs captured in Active Directory). Applications also interact with each other, and the interactions can be captured in the form of Application Programming Interface (API) logs. Many leading cloud providers capture their logs at a more granular API level as opposed to a task level. Often, these logs are used as an audit trail, that is, to capture logs to primarily support debugging issues and to help with investigations. These logs could also be used to help management to achieve their business objectives. This is a largely untapped opportunity waiting to be exploited by companies for gaining insights into the company's behavior, including organizational risks. Understanding and being aware of the company's behavior is the first step towards gaining actionable company-specific knowledge that can be used to predict company behavior trends. Prescriptive insights can then be leveraged to tweak the company's system in such a way that it provides a more effective platform that supports the business strategy. A higher level of collaboration is needed between the three lines of defense in order to take advantage of this opportunity.

Organizational silos are another barrier to the effective management and control of risks. It is easy for the three "lines" to be isolated from each other when looking at the "Lines of Defense" model. For instance, safety inspectors at an organization want to use an IT application to record their inspections on their phones at a remote site. The first line of defense would be the safety operations team that the safety inspectors are part of. The second line could consist of the cybersecurity operations, the compliance team, or the supply chain team, depending on if the solution was created in-house or if it was a commercial off-the-shelf product. The internal auditors would be the third line of defense to independently assess the risks associated with the solution.

In an ideal organization, the operations team needs to engage internal audit and risk management professionals throughout the life cycle of the project, right from the pre-planning phase. Often this is not the case. There is a lack of accountability, communication, and collaboration between the various groups in addressing the risks. Although there may be a centralized risk register, the response is individualized and does not often take into account the overall business objectives. In our example, if the cybersecurity operations team was engaged earlier during the vendor selection phase, the project could have avoided expensive mistakes, like unsupported data formats, additional user access reviews, and redundant processes, etc. On a similar note, if internal audit is not involved earlier in the process, it could result in a reactive approach to dealing with risks introduced by the new solution. The risks would not be assessed until the new solution shows up on internal audit's radar during the next annual risk assessment process.

Now that we understand some of the challenges for managing and controlling risks in an organization, let's look at how the "Three Lines of Defense" model proposes to solve these challenges. We will also look at the recent update to the model, making it the "Three Lines" model. In order for AI and ML to be successfully adopted by internal audit, it needs to be incorporated into the "Three Lines" model.

# The Three Lines Model

The "*Three Lines*" model better clarifies the roles and responsibilities of internal auditors as well as addresses the need for internal auditors to provide advice to achieve business objectives in addition to its existing role of playing defense. This was unclear in the previous model, where management may have been viewed as being solely responsible for helping the business achieve its strategic goals.

Figure 1-2 shows the updated "Three Lines" model. It addresses the need for risk management functions to align with the organization's business objectives. This is a considerable shift from the earlier "Three Lines of Defense" model.

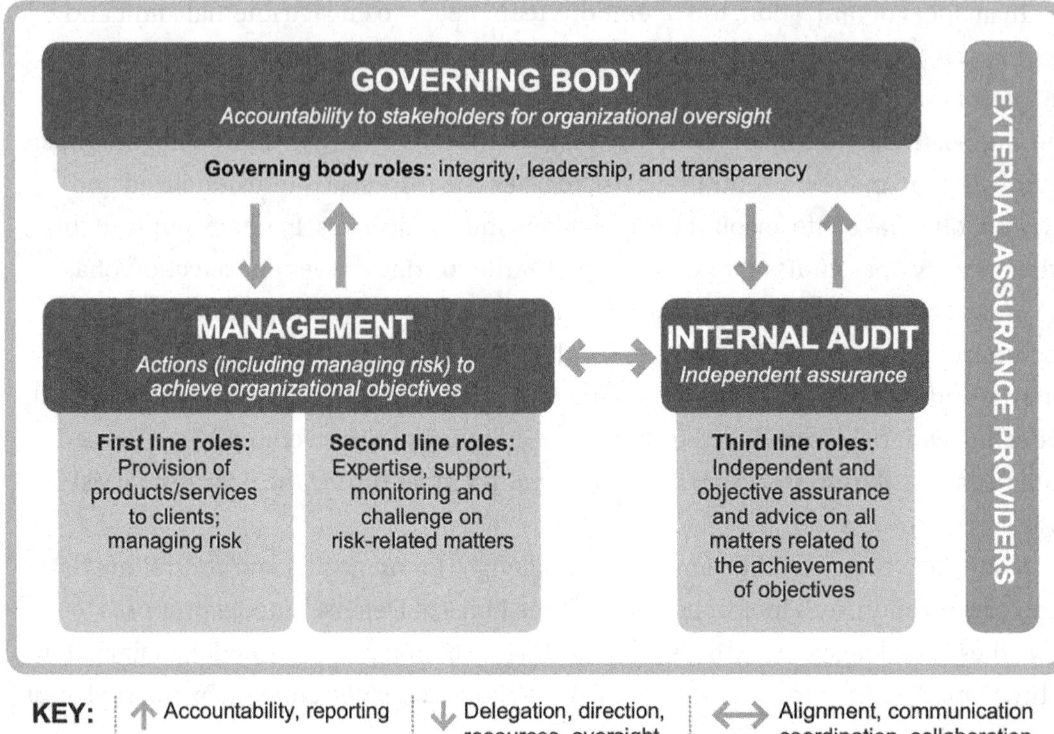


***Figure 1-2.*** *IIA's "Three Lines" model*

The roles illustrated in Figure 1-2 are explored in detail as follows:

- **First line roles.** This type of role involves the management of risk by the providers of products or/and services within and outside the organization.
- **Second line roles.** These roles consist of subject matter experts/ domain experts providing support, monitoring, and assessment of controls for their corresponding functional domain.
- **Third line roles.** This type of role provides independent and objective assurance as well as guidance on the achievement of objectives.

In addition to the definition of the roles and responsibilities, the "Three Lines" model emphasizes the adoption of a principles-based approach to meet organization-specific objectives.

The six principles are stated as follows:

- Principle 1: Governance
- Principle 2: Governing body roles
- Principle 3: Management and first and second line roles
- Principle 4: Third line roles
- Principle 5: Third line independence
- Principle 6: Creating and protecting values

Principle 1 talks about the responsibilities of the governing body, management, and internal audit and how they relate to governance. The roles of each of the lines are defined in detail in Principle 2-4. Principle 5 stresses the importance of the independence of the internal audit function from the responsibilities of management. Principle 6 goes through the creation and protection of business values through collaboration and transparency between the three lines.

# Conclusion

AI/ML needs to work within the "Three Lines" model in order to be leveraged for risk management. The AI/ML application needs to be developed by either the first line or the second line depending on the type of application. For instance, a duplicate vendor payment application will be implemented by the first line since the application falls under the Account Payable (AP) process. An access management review AI application will be produced by the second line because it would be best managed by the cybersecurity team. The oversight and validation of these applications will be provided by internal audit (the third line).

AI and ML pose unique challenges in their usage in risk management applications. The recent change to the "Three Lines of Defense" to clarify roles and responsibilities is a step in the right direction. The "Three Lines" model supports AI and ML applications in the risk management domain by ensuring proper governance structures and processes are maintained. Additional adjustments can be made to this model by further clarifying the AI/ML team's roles. The new model's principles-driven approach is an essential change from the previous model in that it ensures the risk management decisions are consistent. It also ensures that the model is continuously improved over time.

The new "Three Lines" model also emphasizes the evolved internal auditor's role as an enabler for the achievement of the organization's business objectives. The model recognizes that internal audit, along with the other lines, must work together to create and protect business value. The work needs to be aligned with the prioritized interests of the organization. This is in addition to the existing traditional role of internal auditors to help independently assess and validate the organizational risks. AI and ML provide excellent opportunities for internal audits to provide tangible value to the organizations.

# CHAPTER 2

# Common Audit Challenges

The internal audit team performs an assessment of the identified risks. As part of the assessment, process data is used in the form of supporting evidence. For instance, if risks have been identified with the ERP access management process, one supporting documentation could be evidence of access reviews. Access reviews are typically captured as a screenshot from the system showing that management approved the access prior to it being granted to an individual. The assessment can be used to support claims made by interviewees as an independent corroboration tool. It could also be in the form of a review log trail or emails from appropriate authorities showing that they approved the access before the access was granted. The supporting documentation, in this case, can be called *unstructured data*. We will explore more unstructured data in this chapter.

Another usage of data can be in the form of detailed testing. In this scenario, data is obtained as tables with fields. For instance, in the access management audit process, this could include a table showing who has access to what role, a list of all administrators in the system or a list of terminated employees. This type of data is known as *structured data*. We will see more on this in the following section.

Data processing techniques and tools have been continuously evolving over recent years. More efficient ways of transforming and understanding data have been accelerated with the need to monetize data and derive value to the business. The internal audit staff must be able to keep up with this pace if they are to be able to use the organizational data. For example, in the access management audit example, the table showing which employee has access to what role can be pretty significant, may require some joining of data (like tying up employee ID with employee name), and an understanding of the data distribution within the ERP system. The ability of internal audit to perform such data operations depends upon the data literacy of the overall team.



© Maris Sekar 2022
M. Sekar, *Machine Learning for Auditors*, https://doi.org/10.1007/978-1-4842-8051-5_2

Even though some people on the team may be responsible for data-intensive requirements of the group, the importance of such tasks needs to be understood by all members of the internal audit team, from the Chief Audit Executive (CAE) to the junior auditor. The CAE and the internal audit director can better meet the resourcing requirements by hiring the appropriate data specialists or internal auditors with data knowledge for the data-intensive tasks. The managers and the auditors will have a role to play in the adoption of data within the internal audit function. We will explore more on this in the first section of this chapter.

*Citizen development* is a concept that is driven by innovation and one that is gaining more traction within organizations as more people within the organization realize the power of data. What if every person in the organization had the ability to tap into any data produced by the organization? This is the question that citizen development attempts to solve. We will discuss this concept and the challenges associated with governance and security in this chapter. The completeness and accuracy of the data are of great importance to ensure that the quality of data is maintained. A systematic process needs to be developed to ensure that auditors perform essential data quality and completeness checks. Roles and responsibilities for auditors and data specialists need to be designed with great detail to ensure that auditors know when to engage data specialists for more complex tasks. A framework that builds on the previous chapter will be discussed as well.

*Big data* brings its own challenges that have to be addressed for an auditor. The big data concepts of *volume* and *veracity* of data and their challenges will be explored in this chapter. In addition, the idea of *data bias* will be looked at in great detail. This can be one of the more complex problems to detect and overcome due to its complexity. Some of the common types of data bias and the corresponding solutions to overcome them will be proposed here.

# Data Literacy

One of the main obstacles to data enablement (adoption of data) within an audit department is that not everyone is entirely on board with the importance of data. There is a lack of awareness and a coordinated approach at some or all levels of the internal audit team. Data tools and techniques are expected to be the responsibility of the more tech-savvy roles or backgrounds within the team.

Clive Humby coined the phrase "*Data is the new Oil*" in 2006. Raw oil by itself is of less value initially, but when it is converted to other products, such as plastic, chemicals, gas, and minerals, it becomes more valuable as a result. Similarly, raw data by itself is of no value to organizations. Value is derived from the processing and usability of insights gained from the data. This was the original interpretation of the phrase. Building on this interpretation, there is no one role indicating an "Oil Expert" or an "Oil Specialist" within the oil processing steps. A team of people from senior management all the way to the field staff level come together to refine the oil. Why should we think of data to be any different? Data processing must involve the Chief Audit Executive, the audit director, the audit managers as well as the auditors in order to extract the value of data as a team.

Senior audit leadership needs to understand the role of data in empowering the results of the audits. Senior leaders need to provide auditors and managers with the specialized tools and training required to work with data that add value to the organization. The managers' role in extracting value from data is to assess the data analytics knowledge gaps in the team and have a plan in place to address those gaps. Managers need to identify data analytics opportunities that will bring value to the organization. Notice how the focus is on the organization here and not just meeting the objectives of the internal audit team. This builds on the Three Lines model we talked about in the previous chapter that points to the fact that everyone involved in the risk management process must contribute to the organizational objectives.

Not all members of the audit staff need to be trained as data specialists, but a high-level knowledge of all the steps involved in data processing should at least be understood by them. Part of the role of an auditor is to be a project manager. In order to be a good project manager, the high-level tasks of data processing required to ensure that the right processes are being followed for their audits should be understood. As a project manager, an auditor can engage other data specialists to understand the requirements of their audit data needs.

In addition to knowledge about data processing, domain knowledge has to be combined with the gathered data in order to derive the optimal value. Generally defined, domain knowledge is the knowledge about the systems and processes around a specific field or discipline. In the field of internal audit, there are two types of domain knowledge – *audit domain knowledge* and *audit area domain knowledge.* Audit domain knowledge is knowing about the processes within internal audit – such as how to audit and what to audit. Audit area domain knowledge deals with the knowledge about the functional domain of the audit itself. For example, an understanding of the Accounts Payable (AP) process is necessary in order to audit an AP process. When performing audits, it is best to combine the functional domain knowledge (AP process) with the understanding of the data itself (such as vendor payments data) in order to derive the most value out of the audit.

There are many ways to improve data literacy within an organization. Data literacy programs can be developed by leveraging data experts within the organization. Similar to learning a new language, data fluency can be trained with the help of other people who are more fluent with the language of data. It may be more beneficial for learners to learn from data experts who are familiar with the domain area of the learner. Knowledge-sharing workshops, quizzes, and staff presentations are some of the mediums that can be used to promote data literacy. Like any other language, data lingo needs to be practiced on a regular basis so that it can be mastered at a faster pace. Establishing a framework for citizen developers and creating a data center of excellence can help with the practice of data lingo.

One of the main benefits of an organization with a higher number of data-literate employees is the significant improvement in technical communication between the various teams. Moreover, a deeper level of understanding of the organizational data is realized, which can lead to other benefits, such as unlocking hidden insights within the data.

## Manual Testing

In the traditional audit testing process, there are many sampling methods to choose from. *Random sampling* is commonly used for its effectiveness among the other sampling methods. It consists of the selection of random samples based on the level of risk (low, medium, or high) involved in the audited process and the number of transactions generated by the audited system. This is the non-statistical way of sampling, but in essence, both the statistical and non-statistical sampling methods involve picking a subset of transactions to look at. As an example, suppose we have an access requests review audit with a high level of risk based on the risk assessment. The access management tool/system generates about 50 access requests a day. Based on these two variables, the level of risk and the number of transactions involved, we would typically refer to a sampling table that indicates the number of samples to use. Let's say that the sample size is 40. During the audit testing process, a sample of 40 will be randomly selected from the total access requests pool. It is up to the auditor to use the appropriate randomization process.

There are many problems associated with the preceding process. Only a subset of the data provided by management is tested for audit processes. Often this is a small percentage of the total data available for testing. Testing of 100% of the population was traditionally carried out only in rare circumstances. A rare circumstance includes a high-risk process or if the control had failed at a previous time. For example, if you were looking at examining vendor payments made by the organization, you would test all vendor payments if there were multiple instances of fraud committed in previous years. This type of testing, where 100% of the population is tested, is called a *test of details*. There is no set guideline on how this should take place. Should all the population should be tested if the total population size is only 50? It is up to the auditor's judgment to choose the appropriate method. It could take many years of practice before an auditor is able to pick a proper testing method.

Random sampling does not necessarily mean the samples will be tested manually by the auditor, but often this is the case. After random samples have been picked, the auditor typically retrieves the original documents and compares them manually to the numbers from the system being tested. This can be a time-consuming process. Even though only a sample of the population is tested, it can take some time based on the testing scope. Suppose the payments are being reconciled to invoices as part of a vendor payments audit. The auditor will retrieve 25 sets of data from management or directly from the system. Each set will contain one invoice and the payments that were made as part of that invoice.

Figure 2-1 shows how audit performs testing for reconciling invoices to payments. Figure 2-1 represents one test out of a series of tests that need to be carried out for the vendor payments audit.

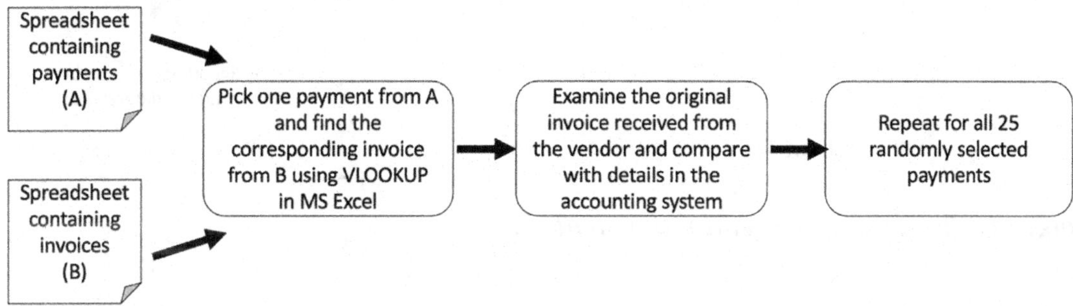


***Figure 2-1.*** *Manual vendor payments audit*

The time spent on doing manual testing could potentially be used for other audit tasks. There is also uncertainty involved with manual testing due to human error. If the auditor switches up the invoice or the payment due to fatigue when testing, this may cause additional time to be lost in the process.

*Automated audit testing* involves the use of scripts to automate tasks (such as audit testing). For instance, the vendor payments testing discussed earlier can be programmed as a script to reconcile the payments to their corresponding invoices. Any invoices involving a difference can be noted as an exception and the auditor can follow up with management. Moreover, this testing can be extended to 100% of the population. If the tests work for the random samples, they should work for 100% of the population. Another advantage to automated audit testing is the ability to reproduce the results. Since the testing steps are captured in a script, it is easy to replicate and use for other similar audit tests. The vendor payments scripts can be leveraged for the accounts payable test with a slight modification. This a significant win for the organization and it can help the organization optimize its audit resources as well as meet its business objectives sooner.

Figure 2-2 shows one possible setup of the automated audit testing process for the vendor payments audit.

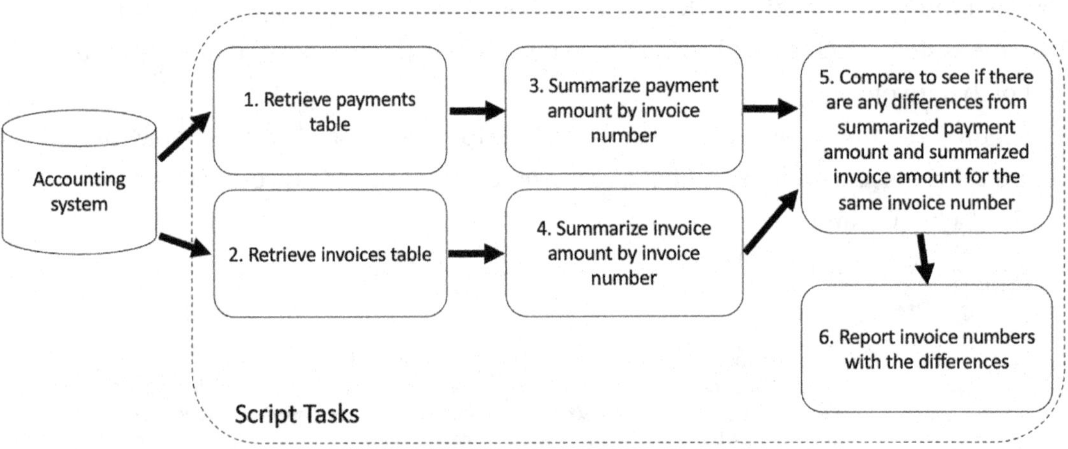


***Figure 2-2.*** *Automated vendor payments audit*

## Data Sources

Many companies have multiple data sources supported within their technology ecosystem. This could be due to a number of integrations with their business partners such as vendors or customers. In order to be effective at identifying and mitigating risks, auditors need to be aware of the various data sources and the general organization of data. As an example, during the risk assessment step for looking at the Accounts Receivable process, there is a need to gather all systems and their corresponding data in order to identify the risks associated with them. If we fail to understand the risks associated with a crucial data repository that maintains the list of customers that owe money to the organization, this can be a problem. Hence, all data related to the process being audited need to be included during the risk assessment step. The process owner plays a critical step in ensuring the auditors are aware of the different data sources of the process. But, in some cases, it may not be clear what data sources are involved due to the complexity of the system. In order to overcome this hurdle, experts with the domain area knowledge can be utilized to check if all data sources have been identified.

After the identification of data sources, the next step is to understand the organization of data. Although a deep level of understanding of all data sources may not be required, a general knowledge of them is important. To continue from our Accounts Receivable process, the list of customers that owe the organization money may need to be retrieved by joining with multiple tables. Data may be organized in such a way that Table A may contain the amounts customers owe the company along with the customer number and last payment date. Another table, Table B, may include the customer number and the customer details such as customer name, address, and other contact information.

Figure 2-3 shows a sample case of how the Accounts Receivable process data may be organized within an organization.

Table A

| Customer # | Payment Amount | Last Payment Date | ... |
|---|---|---|---|
| 120123 | 1000 | 10/01/2019 | |
| 120124 | 5400 | 05/02/2018 | |
| 120125 | 9800 | 11/05/2020 | |
| ... | ... | ... | |

Table B

| Customer # | Customer Name | Address | ... |
|---|---|---|---|
| 120123 | ABC Tech In. | 123 Henry W. | |
| 120124 | XYZ Inc. | 9 Potter Dr. | |
| 120125 | QRS Corp. | 18 Edge Way. | |
| ... | ... | ... | |

| Customer # | Payment Amount | Last Payment Date | Customer Name | Address | ... |
|---|---|---|---|---|---|
| 120123 | 1000 | 10/01/2019 | ABC Tech In. | 123 Henry W. | |
| 120124 | 5400 | 05/02/2018 | XYZ Inc. | 9 Potter Dr. | |
| 120125 | 9800 | 11/05/2020 | QRS Corp. | 18 Edge Way. | |
| ... | ... | ... | ... | ... | |

***Figure 2-3.*** *Accounts receivable process data*

In order to know the customers that are overdue in payments to the company, Table A and Table B will need to be joined to produce a single combined table that contains the amounts owed, the last payment date, customer name, address, and other contact information. Remember that this is just one way that data may be organized. Moreover, during the course of the Accounts Receivable process audit, we may need to get additional information to check this data.

In order to check the last payment date in Table A, an auditor might choose yet another table, Table C, that contains all the payments received from the customer. Table C will need to be queried to get the latest payment date, and this can be used to check with Table A's last payment date to support the audit.

Figure 2-4 demonstrates how Table C may be used to check the latest payment date to compare with the latest payment date from Table A.

Table A

| Customer # | Payment Amount | Last Payment Date | ... |
|---|---|---|---|
| 120123 | 1000 | 10/01/2019 | |
| 120124 | 5200 | 05/02/2018 | |
| 120125 | 9800 | 11/05/2020 | |
| ... | ... | ... | |

Table C

| Customer # | Payment Amount | Date | ... |
|---|---|---|---|
| 120123 | 350 | 06/04/2019 | |
| 120124 | 5400 | 05/02/2018 | |
| 120123 | 650 | 10/01/2019 | |
| ... | ... | ... | |

| Customer # | Payment Amount (A) | Payment Amount (B) | Difference | Exception | ... |
|---|---|---|---|---|---|
| 120123 | 1000 | 1000 | 0 | No | |
| 120124 | 5200 | 5400 | 200 | Yes | |
| 120125 | 9800 | 9800 | 0 | No | |
| ... | ... | ... | ... | ... | |

***Figure 2-4.*** *Accounts receivable reconciliation*

# Structured vs. Unstructured Data

Data comes in many forms and types. An auditor may need to differentiate between the different types of data in order to support AI/ML applications or initiatives within the organization from a risk management point of view. Generally, they are categorized into two distinct types - structured data and unstructured data. Knowing what kind of data is being dealt with is key to determining the techniques used to extract, transform, and load (ETL) along with the effort required to apply the data towards business outcomes.

Structured data is well-defined and is stored as columns and rows that can be easily searched or analyzed. For instance, it could contain information such as a list of users that have access to the system. The data is stored in such a way that each row contains information about one user in most cases. The columns contain attributes about the user. The table looks similar to an MS Excel Spreadsheet with rows and columns. A particular user's information can be retrieved by searching for the user in the username or user id field and then reading across to get the corresponding information for the user under question. Attributes could be Name, Address, Contact number, etc.

Structured data is the most dominant format used in AI/ML applications currently, but this can change. More on this is covered later. It is widespread in ML applications because most of the algorithms perform best with tabular data. Since each user (or observation) is stored in exactly one row and each row contains the attributes (or features) of that user, it is easier for most ML algorithms to "learn" from this simple convention. In order to ask the ML algorithm a question about the user, the data used to learn must be in the format specified earlier. There may be more advanced ML algorithms that summarize the user information first and then train the model using the summarized data. This will be explained in detail in the chapter dedicated to machine learning.

See a user list that contains structured data in Figure 2-5. Every line in the table represents a single user and their corresponding user information.

| User ID | Name | Address | Phone | Email | ... |
|---|---|---|---|---|---|
| 120123 | Charles, W. | 123 Henry W. | (123)-1911212 | wcharles@abcinc.ca | |
| 120124 | Smith, A. | 9 Potter Dr. | (555)-2321200 | asmith@abcinc.ca | |
| 120125 | Cora, L. | 18 Edge Way. | (999)-3245342 | lcora@abcinc.ca | |
| ... | ... | ... | ... | ... | |

***Figure 2-5.*** *Structured data – user list*

Unstructured data represents data such as text that is not stored in a well-organized manner. It can be harder to search or analyze the data. Humans are generally better at understanding unstructured data when compared to AI/ML algorithms. Although it is easier for humans to analyze, it is not scalable. Think of a human reading a book and understanding the content. The text in the book represents unstructured data. In order to get the information in the book, the human has to read through the text and interpret the meaning of the text. So, it requires some cognitive power. A human is limited to how much they can read. If an AL system was programmed to read the text and gather information from the book, it might be scalable, and most likely, the AI system would be able to read more books than a human could.

The problem is designing a robust AI/ML system that can interpret unstructured data. Although it is a more challenging problem to solve, there are packages available in major programming languages that now readily support Natural Language Processing (NLP). You can use the NLP libraries to make an AI agent that reads books in the same genre and writes an entirely new book based on the read books. Another common application of unstructured data is in the form of user reviews. Imagine if you could read Amazon reviews of products and find out if the total sentiment of the purchasers is positive or negative based on analyzing the data.

Figure 2-6 shows an example of unstructured data. It captures user reviews in plain text. Every two lines contain the details of the user and their assessment of the product.

User A reviewed Product #123:
A great product! Astounding customer service.

User B reviewed Product #324:
I like the product. Would have liked to see feature A.

User C reviewed Product #987:
Absolutely fantastic! Will definitely buy again.

...

***Figure 2-6.*** *Unstructured data - user reviews*

From an auditor's point of view, the challenges of structured and unstructured data need to be understood. This is crucial since all the risks associated with their use in AI and ML applications need to be covered. Most risks (such as survivor bias, expert bias, and data integrity) are similar between structured and unstructured data. The challenges associated with data integrity and how to overcome them were discussed in Chapter 1. Data Bias will be discussed in detail later. Unstructured data poses some additional challenges due to the complexity of the NLP mechanism. For instance, when reading a review from Amazon, it can be observed that in certain situations, the wrong sentiment may be recorded.

Consider this Amazon user review: "I am *not un*opposed to the product." Some less sophisticated NLP algorithms may mark this review as being negative when in reality, the user may have had a neutral sentiment. A review may contain formatting inconsistencies such as extra breaks and spaces. Additional data transformation or cleanup steps may be required before the NLP algorithm is run on it if the NLP algorithm does not support this functionality natively.

# Citizen Developers

Data democratization is becoming more popular, especially with the push for gaining data-driven insights within processes. Historically, one of the main barriers for working on data analytics projects, as in any other projects, has been based on if the Information Technology (IT) team had the right technical talent and resources to work on projects.

Often, organizations prioritize projects due to the constraint on resources and specialized people. The concept of citizen developers alleviates this problem by enabling non-technical individuals within the organization intuitive access to data, tools that are easy to use, and with the ability to automate manual tasks. Simply defined, citizen developers are non-technical individuals within an organization who use company-supported tools that are easy to use.

As an example, consider the requirement for an Accounts Receivable (AR) team to develop an analytics dashboard to analyze payments that are owed to the company. After planning work is done, there are two ways to get tools and people to work on this project. The first way is to utilize a knowledgeable AR team member to develop the dashboard. The second way is to engage the IT team of the organization, which is usually the cheaper route. If the AR team were to utilize someone knowledgeable with data analytics, it would be ideal to ensure the domain knowledge (about the AR process) is leveraged during the development process. With both ways, the time taken to create the dashboard may vary but is often based on the learning curve of the technical people and their availability.

Imagine if the tool used to develop the dashboard is easy to use and has a drag-and-drop Integrated Development Environment (IDE) interface. In this case, the AR team may be trained with this easy-to-use tool along with other teams in the organization. With this scenario, the AR team is able to leverage their domain expertise to develop a dashboard that is fit for the purpose with less chance of misalignment with the project requirements.

The organization can leverage the technical resources in the citizen developer setup by ensuring proper training and guidance from a governance point of view (if it is available). In our case, this means sharing company-led training and ensuring data is restricted only to authorized developers (not all employees require access to Accounts Receivable data). The organization should also provide easy-to-use tools such as Qlik, Tableau, Smartfire, and Power BI along with support for these tools. The organization may also want to develop templates and general guidelines to ensure that all company dashboards are standardized and easy to understand if another team were to leverage the same dashboard. The audit team can access the dashboard to perform an audit on the AR process. Another example use case would be for senior leaders to have visibility over the AR process by directly accessing and exploring the data to get an understanding.

One of the main challenges of the citizen develop setup is ensuring that access to data is restricted to only authorized members of the organization. One way to ensure that the information is secure is by ensuring only the authorized team has access to the data facilitated by permission management on the data warehouse. Most data warehouses support security features that enable permissions to be controlled at various levels - individual, team, organization, etc.

A data warehouse is a snapshot of the live database that is taken on a regular basis. The snapshots can be scheduled to be taken monthly, weekly, daily, or hourly, depending on the consumers of the data. The data warehouse concept helps address another problem associated with live production data. A company makes copies of the live production data and the application code to ensure that any change is appropriately reviewed and approved by the relevant individuals before the change is applied to production data. This is crucial in order to avoid interference with valuable live production data.

The diagram in Figure 2-7 illustrates the data warehouse and IDE setup for a sample use case of the AR Process.

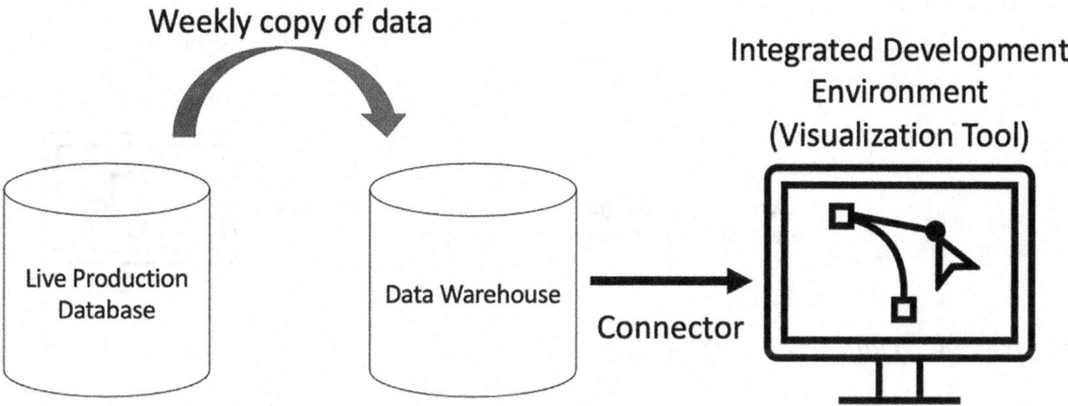


***Figure 2-7.*** *Sample data warehouse and visualization tool setup*

Data quality is an important point to consider when allowing citizen developers access to raw data from a risk management point of view. There are two possible setups that are normally observed for taking care of data quality. In the first setup, the "scrubbing" or cleaning of data occurs before data arrives in the data warehouse. In the second setup, raw data is deposited in the data warehouse without any cleaning procedures. In the case of the first setup, data quality is taken care of, because it arrives from clean data sources or the data management team "scrubs" the data centrally for the citizen developers. In the second setup (which is the more common setup), data will need to be cleaned possibly with the drag-and-drop IDE or with some other data cleanup tool before being loaded into the IDE. No matter what type of setup is used, data must be of the best quality possible before it is available to the end-user whether the user is a business analyst, manager, director, or member of senior leadership.

Figure 2-8 shows the two different setups that are normally seen implemented. The second setup is the most commonly seen setup.

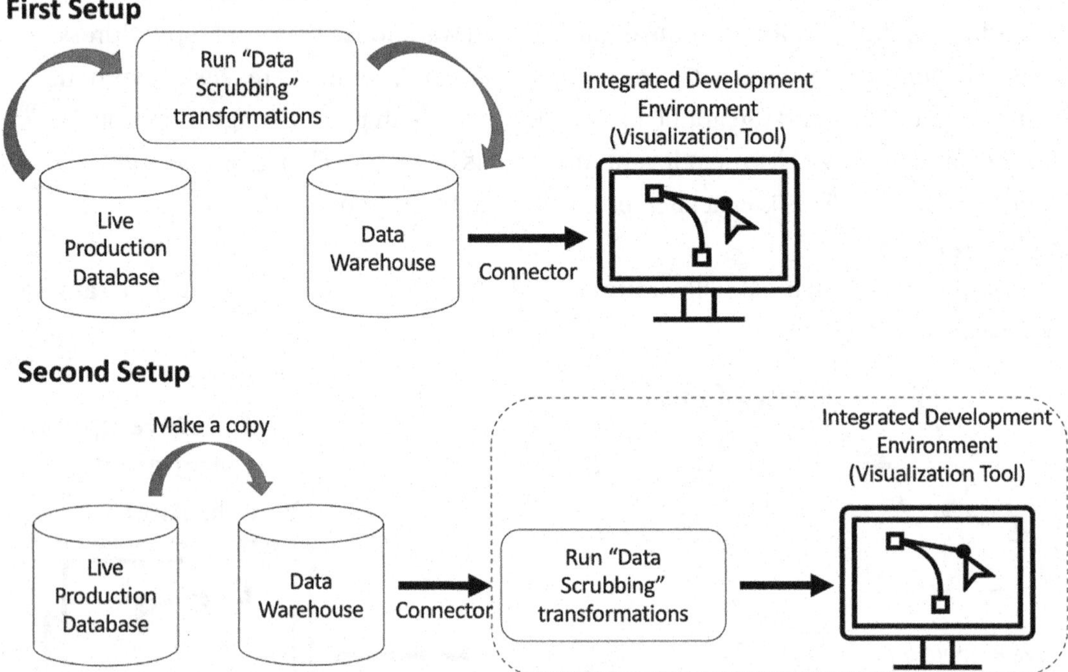


*Figure 2-8. "Data scrubbing" setups*

Automation of manual tasks using a simple to use drag-and-drop IDE is also becoming more popular with the advent of Robotics Process Automation (RPA). RPA was initially intended for technical developers, but the current trend is to make it available across the organization even to non-technical citizen developers.

In RPA, automation scripts or recorded steps mimic the behavior of an end-user. This can be especially useful if the user behavior is a recurring action that needs to take place on a regular basis. For example, an end-user may have manually processed paper invoices received from vendors by copying the invoice details into the ERP system. For simplicity, assume there are two network directories to help with this process – "Received Invoices" and "Processed Invoices." The received invoices are deposited into the "Received Invoices" directory, and after processing, each processed invoice is moved to the "Processed Invoices" directory. The steps can be broken down as follows:

1. If an invoice exists in the "Received Invoices" directory, read its contents.
2. Validate if the Purchase Order exists in the ERP system and if the amount being invoiced is within the available amount in the corresponding PO in the ERP system.
3. Open the invoice entry screen in the ERP system.
4. Populate the fields in the invoice entry screen with the read invoice.
5. Perform validation check using the ERP's validation functionality.
6. Submit the invoice for issuing the payment to the vendor.
7. Move the processed invoice to the "Processed Invoices" folder.

An RPA tool can be utilized to capture the preceding steps either programmatically or via a user-friendly drag-and-drop IDE. The steps represent a breakdown of the high-level tasks involved in processing an invoice.

For step 1, an Optical Character Recognition (OCR) tool may be applied if the received invoice is an image. Some RPA tools have pre-built functionality that supports them. If the functionality is unavailable, another tool may need to be leveraged to first convert the image documents to readable text.

Steps 2 and 5 are essential steps to ensure that data quality is preserved throughout the automation process. Often the ERP system might provide some sort of validation before the entered data is processed. It may be useful to leverage this validation check as an added check. The SAP ERP system provides it for most of its standard entry screens. It is necessary to ensure the processed invoice is moved to the "Processed Invoices" folder as a way to track if the invoice was processed successfully. The RPA tool may provide logs to record the tasks along with their corresponding timestamps. These RPA-generated logs can be used as an audit trail and can be leveraged when conducting an audit.

Access management is not an issue for de-centralized RPA tools that reside on the user's machine and uses only the access that is already available. If the RPA tool is a centralized application that is to be used by technical specialists, then access management poses many challenges for auditors. From a citizen developer's point of view, de-centralized RPA tools that use only the access provisioned for the user is the ideal solution.

## Data Wrangling

In order to answer a business question, raw data needs to be gathered, filtered, transformed, and loaded into the analytics tool. This process of extracting, transforming, and loading data is called data wrangling, data cleaning, or data munging. As mentioned earlier in the data literacy section, data by itself is of less value before it is processed in some way. When data is processed, specific information that adds value to the organization is extracted.

Data wrangling is a time-consuming process that may take up to 80% of data analytics projects. Even though auditors do not need to perform data wrangling by themselves, they do not need to understand how the raw data was transformed into the final result. From a risk management point of view, it is important for auditors to understand the risks (introduced or eliminated) due to the processing of the raw data.

Consider a case when auditing a dashboard used by the CXO suite to make investment decisions based on organizational data. Let us assume the raw data contained multiple currencies for the Return on Investment (ROI) due to the business operating in multiple countries. If the final numbers shown in the dashboard did not take this into account and no one checked for this, the decisions made by the CXO suite may be influenced to make the wrong decisions. Although this scenario may seem exaggerated, it demonstrates the value of knowing how data gets processed. So, an auditor needs to review how data was transformed just before it was loaded into the dashboard to ensure that a problem like this doesn't occur.

In the case of machine learning applications, data munging is almost always performed at the beginning of the problem solving process to make data consumable by the machine learning model. After predictions are made, the conclusions may need to be interpreted before the final result of the predictions is displayed. More details on this will be explored in later chapters. If auditors were to look at machine learning applications, they would need to look at how data was transformed at these stages of the machine learning development cycle.

One of the main challenges of data munging is that the data transformations can be hard to follow, especially for non-technical auditors. In this case, other technical auditors or subject matter experts (SME) of the process may be used. It may be useful to map out the transformations and data sources. A data lineage diagram can be a useful tool in analyzing the transformation to the data and how data flows to the final result.

# Data Bias

Data bias is a common problem with all types of data analysis. Bias is an interference to the actual data that causes prejudiced, preconceived, or predetermined notions to be embedded within data.

For instance, imagine that a company was to design a Bluetooth-enabled dog collar. A decision needs to be made by the designers on the maximum circumference that the collar will support. Assume that the company wants to use data to power their decision. They conduct an online survey to ask their potential customers the circumference of their dog's neck. When the company analyzes the data, they find that most customers reported 40 cms and 50 cms, which represented about 80% of the participants. The other 20% of customers reported a value that was between 35 and 55 cms (for example, 35.5, 36.3, 40.2, etc.). Since 40 cms and 50 cms appear to cover most of the customers, they make a decision to make collar sizes of 40 cms and 50 cms.

Unfortunately, the company did not end up selling nearly as much as they thought the demand had been. Upon probing and revisiting the survey results again, they came to an insightful conclusion that participants could have potentially selected 40 cms and 50 cms because there was something else at play. The survey had three possible options - 40 cms, 50 cms, and "Other, please specify," where customers can enter the number themselves. It was easier to select one of the options with a value than to select "Other" and specify a value for it. This explains the bias towards 40 cms and 50 cms.

Figure 2-9 illustrates the survey set up to get the dog's neck circumference from survey participants.

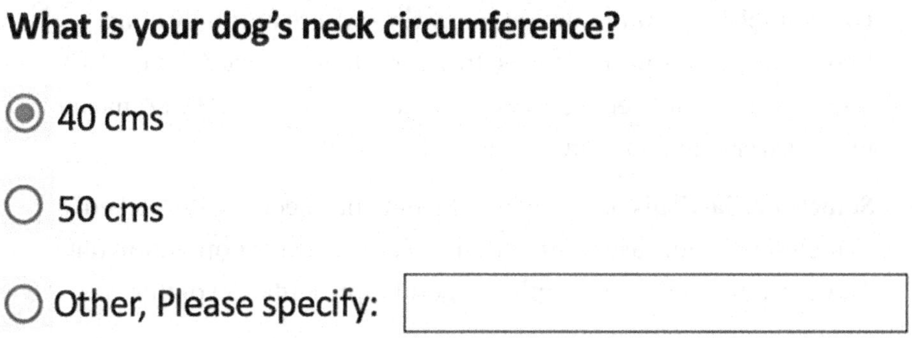

**What is your dog's neck circumference?**

40 cms

50 cms

Other, Please specify:

***Figure 2-9.*** *Online survey to get the neck circumference*

Now consider the same preceding example, but with a different type of survey. What if the response in this new survey was a free text field? Would that have improved the quality of the survey results? The results may be better because the survey participant is forced to enter a value. There is still a chance of data bias in this type of survey. It is known that people generally prefer whole numbers, so if the circumference was measured as 49.3 cms, it might be rounded down to 49 cms, and if the circumference was measured as 49.7 cms, it might be rounded up as 50 cms.

Bias can also occur based on how data is interpreted, that is, due to the preconceived notions of the interpreter. This is best explained by the proverbial phrase, “Is the glass half-empty or half full?” Based on if the interpreter is optimistic or pessimistic, the answer would be different. Although the fact remains that the glass is half-filled with water in both cases, it can be interpreted to mean different things.

Machine learning uses some form of historical information to make decisions or predict the next move. It is important to consider data bias for machine learning applications because data bias can erode the accuracy of data. As data gets eroded, it can get to a point where the data misrepresents itself and does not reflect the population data anymore. This can cause decisions to be inaccurate. There are different types of data bias when it comes to machine learning applications. As an auditor, it is essential to know about these types of data bias in order to leverage machine learning applications for risk management. Also, when auditing machine learning applications, these types of biases need to be evaluated to ensure the data is accurately represented.

Here are some of the different types of data bias:

- **Activity Bias.** Many data points are generated from a few users. For example, it is common that many reviews are normally done frequently by a handful of users. In an article by Baeza-Yates in 2018 titled “Bias on the Web,” it was stated that more than 50% of Amazon reviews were generated by 4% of users.

- **Selection Bias.** This occurs when the data has been “hand-picked” to include certain cases and does not reflect the real population data. The preceding survey example is a good representation of this.

- **Model Bias.** It occurs if the machine learning model is overfitting (when the model is overcomplicated due to many different types of data points) or underfit (when the model is too simplistic due to a smaller number of data points).

- **Societal Bias.** This bias is caused due to training a machine learning model with data generated by humans. For example, training a machine learning model with news articles from a news agency that is funded by a political organization. The decisions made by the model could potentially favor the political organization in the long run because the content is biased towards the political organization.

# Conclusion

In this chapter, we looked at some of the common challenges faced by auditors in the form of people, process, and technology. From a people point of view, it is important to consider the level of knowledge of data within all levels of the Internal Audit team. Another aspect of people is Citizen Developers within the organization who might need additional support to help them achieve their goals. In terms of process, manual testing, data wrangling and data bias pose challenges that are unique to handling data in general and data sampling. Lastly, technology challenges discussed include issues related to big data in the form of volume, veracity, and types of data. AI solutions must be able to overcome these challenges in order to successfully implement audit applications.

# CHAPTER 3

# Existing Solutions

This chapter will look at some of the current solutions available in the space of auditing. The solutions mentioned here are by no means a complete listing, but they are some of the most commonly used technologies in the field of risk management. The solutions include substantive testing, CAATs, process mining, and continuous auditing.

Substantive testing is used to ensure the numbers stated in financial statements are accurate. Substantive testing or test of details use 100% of the data to ensure numbers balance. This will be discussed in detail in this chapter.

Computer Assisted Auditing Techniques (CAATs) are used to describe a wide array of tools and techniques used to automate IT audit processes. The common application of CAATs is in the reconciliation between General Ledger and the Trial Balance. Some technologies are built to meet a particular audit need. These fit-for-purpose technologies may help organizations identify suspicious vendor payments or fraudulent expenses. In many of these technologies, proprietary tools are built for a particular IT auditing purpose for an organization. It could also be used to assess technologies and support investigations.

Process mining is a growing area of interest in the field of risk assurance. In process mining, event data is processed to get insights and actions into workflows. It helps understand process events from an architectural point of view. This will be explored in detail later. In continuous auditing, high-risk audit tests can be automated and run throughout the year. As exceptions are encountered, they are escalated to the appropriate parties so that they can be actioned on.

## Substantive Testing

Financial statements are formal written records that capture the financial activities of the business along with other details about the business. Financial information is produced in a level of detail that is easy to understand. As part of substantive testing, auditors can



© Maris Sekar 2022
M. Sekar, *Machine Learning for Auditors*, https://doi.org/10.1007/978-1-4842-8051-5_3

be involved in testing for errors in individual transactions or fraudulent transactions. An important point to note is that 100% of the test population will need to be tested so that the completeness, validity, and accuracy of the financial transactions can be confirmed. Historically, spreadsheet software (like MS Excel) has been used to run these tests. As the number of transactions becomes larger, so does the complexity in running tests for details. For large businesses, millions of transactions that were observed throughout the year may need to be tested using substantive testing. In order to analyze millions of transactions, specialized audit tools other than spreadsheet software may be required. One popular audit software that can help with this is Audit Command Language (ACL or more recently known as Galvanize). A specialized audit software has the ability to process and store millions of transactions efficiently. They also have a robust audit trail management built into the tool in order to record the operations performed on data.

Even with specialized audit software, the large size of transactions can be a daunting problem. Some tests could take up several hours at a time to complete. The most complex operations include joining multiple tables together and summarizing the transactions at the account level for comparison. CAATs may be used for substantive testing. More details with an example are given in the following section.

## CAATs

CAATs stands for Computer Assisted Audit Techniques. It covers a wide group of tools and techniques used to support auditors in performing their verification. Every process has a risk associated with it. To reduce or mitigate risks, controls need to be put in place. The Institute of Internal Auditors (IIA) refers to CAATs as "the use of technology to help you evaluate controls by extracting and examining relevant data."

At the heart of CAATs, scripts are employed to run the steps auditors would have to perform to evaluate controls. Otherwise, auditors would have to perform these steps manually. For instance, consider the case of substantive testing using CAATs. The general ledger and trial balance over the selected period are obtained and compared to ensure they reconcile at the account level. A General Ledger contains a company's accounting data and is used to record all the transactions that are posted from journals, accounts payable, accounts receivable, cash management, fixed assets, purchasing, and project. A trial balance contains all the debits and credits made to the account during a specified reporting period. The following steps indicate the standard process to reconcile the general ledger and trial balance:

1. Filter the general ledger for the reporting period.
2. Create two columns - one for recording debits and another for capturing credits.
3. Summarize the debits and credits general ledger at the account level.
4. Join the summarized table from step 3 to the trial balance obtained for the same reporting period based on the same account.
5. Compare the debits and credits between the general ledger and the trial balance from joined table in step 4.
6. If there is an account that does not balance from step 5, report it as an exception for further examination.

The reporting period could be a year or less than a year (interim). For step 2, the general ledger amount is recorded as a debit if the amount is negative, or as a credit if the amount is positive. For step 3, the debits and credits for the same account name are added together as part of the summarization process. When joining the summarized general ledger table to the trial balance table, the account ID will be used as the unique identifier. At the end of step 6, all the accounts that do not reconcile are reported along with their details.

The preceding steps can be captured into a script. The advantage to having it in a script is that the script needs to be developed only once. After verifying that the script works as expected, it can be run again for future reporting periods by switching out the data tables with the new data. CAATs have been slowly replaced with data analytics. Data analytics is more focused on deriving value from data through visualizations for the organization. Many of the processes used for CAATs may still need to be followed by a data analytics project, but generally, a data analytics project produces a data visualization that helps the business make decisions.

# Fit-for-Purpose Technologies

In the previous section, we talked about CAATs, which use automation to verify if the controls are functioning effectively. In this section, we will look at some fit-for-purpose technologies that may not be part of an already existing control. Fit-for-purpose solutions can be used to address a specific risk to the organization. For example, a fit-for-purpose solution can be used to counter a widespread problem with vendor payments for an organization. There could be many reasons for such a problem:

- Many vendors of the company are numbered corporations. Numbered corporations are operated by smaller businesses. There may be a lack of established controls in place to prevent fraud.
- The current controls for the accounts payable process may be weak. For example, the training for the accounts payable process may not be detailed enough to prevent fraud.
- An employee within the organization may conspire with a vendor to commit fraud.
- The current accounts payable process may be complex, and there are chances of errors being made for vendor payments.

In order to overcome the problem with the vendor payments process, a test can be built to check if unintended or fraudulent payments were made to the company's vendors. Here are the sample test steps to mitigate the risk from the vendor payments process:

1. Extract all payments made out to vendors along with the corresponding invoices and purchase orders from the accounting system during the test period.
2. Check to see if the payments agree with the invoices and purchase orders. If the payment amount is considerably larger than what's listed in the invoices and purchase orders, mark the related transactions as an exception.
3. Do a fuzzy match of the vendor's name with the company's vendor master list to see if there are any other vendors with similar names. If there is a match, mark the vendor as an exception.
4. See if there are any reversals within the accounting system for the payment in question. Mark these payments in another list to investigate further.

5. Check to see if there is a vendor's contact and banking information that was recently changed. Mark these vendors and the payments associated with them as an exception.

6. Validate the payment amount. Identify recurring payments, ensure the amounts match. If there was a change in amount - mark them as an exception.

For step 1, the corresponding invoices and purchase orders will need also to be extracted along with the payments. This can be normally done through fields on the payment that point to the associated invoice and purchase order. The invoices and purchase orders need to be extracted to ensure the payment amounts line with the invoices and purchase orders. If there is a discrepancy, for example, if the payment amount exceeds the invoice amount, it may need to be investigated further.

For almost all the preceding steps, we capture the related transactions as an exception. This is easier said than done. We need a way to capture the payment that caused the exception along with the info needed to investigate further. We may also need to capture the reason for the exception so that if there are a lot of exceptions, it is easier to group and investigate the exception payments together.

For step 3, the vendor's name can be matched "loosely" with the vendor master list. This is done to ensure the correct vendor details are being attached to the payment under question. For instance, if a vendor's name is "ABC technologies inc.," a fuzzy match would pick up "AB technologies inc.," "ABC technologies," ABC inc.," "abc technologies inc.," etc. This is important because if two vendors with similar names exist within the system, there is a chance that a wrong vendor detail gets attached to the payment. The consequences can be detrimental to the organization. In this case, a wrong vendor could be paid. One may argue this is rare and does not normally happen. The fact of the matter is when the finance team attaches the vendor details to a payment manually, they may be time-constrained and may not notice the correct vendor during the assignment. Manual processes often have human touchpoints that could introduce error into the process.

For step 4, all reversals related to the payment should be checked if the accounting system links all of them together. Reversals can be used as a way to discover suspicious or unintended activity. It also indicates if there is a break in the accounting process. For instance, if multiple payments are discovered every month that have been reversed and paid again, it could mean there is a systematic issue with the payment process. An analysis may be performed to confirm this, and then the process owner can be engaged to work on a viable solution for this process.

Step 5 is a crucial test, but often this is not implemented in many organizations due to the complexity involved and the coordination needed between multiple teams. It checks to see if the contact information or financial information for the vendor recently changed. One of the common ways for fraud to occur is to change contact or banking information. Normally, when a change to the contact or banking information is notified by the vendor, controls are in place to check if the new contact or banking information is valid. This may be harder to check with smaller companies that have a maximum of two employees, especially if one of the previously trusted employees is committing fraud. From a test implementation point of view, this is an easy check to see if the contact or banking info was recently changed from the vendor master list.

In the case of step 6, recurring payments can be identified and checked to see if the payment amount changed over time. Simple logic to identify recurring payments could be to check if the three consecutive payments with the same vendor and amount are being made. Once a recurring payment is identified, the amount that is normally recurring can be compared with the payment amount from the current test period. If there is a discrepancy between these amounts, the identified payments can be explored further.

The preceding test steps are intended to be run over a test period (quarterly, monthly, weekly, etc.). This is by no means an exhaustive list of all validation checks to put in place, but it's a great place to start. One important point to note here is that no matter what exceptions are being investigated, the business needs to be engaged to confirm if fraud or unintended payment did indeed take place. Ultimately, it is the business that is the domain expert and is the most knowledgeable with the process. Just because there are exceptions doesn't mean they are actual problems. Careful consideration and probing need to take place before the exceptions are escalated or reported. This should include, but not be limited to, talks with business and functional experts to confirm the exceptions.

# Process Mining

Almost all applications in the Information Technology (IT) landscape capture an audit trail within the application. Most of these logs are seldom used to derive insights or an understanding of the process the application supports. Process mining is the application of data analytics to event logs to derive actionable insights into process management.

For instance, consider an application that supports a workflow to request vacation days as part of a standardized Human Resources (HR) process. Suppose the workflow is set up in such a way that the employee or contractor can fill up an online form. In the online form, the employee or contractor can enter the details about the vacation request (like the days they are off, who their manager is, etc.). Upon submitting the form, the form is routed to the supervisor or manager for approval. When the manager approves the request, it is routed to the HR manager to do a final check. The manager can choose not to approve the request, in which case the request may go back to the employee to revise the submission. If the HR manager does not approve the request, it is routed back to the employee to revise the information.

Figure 3-1 shows how a sample vacation requisition workflow may be set up as part of the company's HR process.

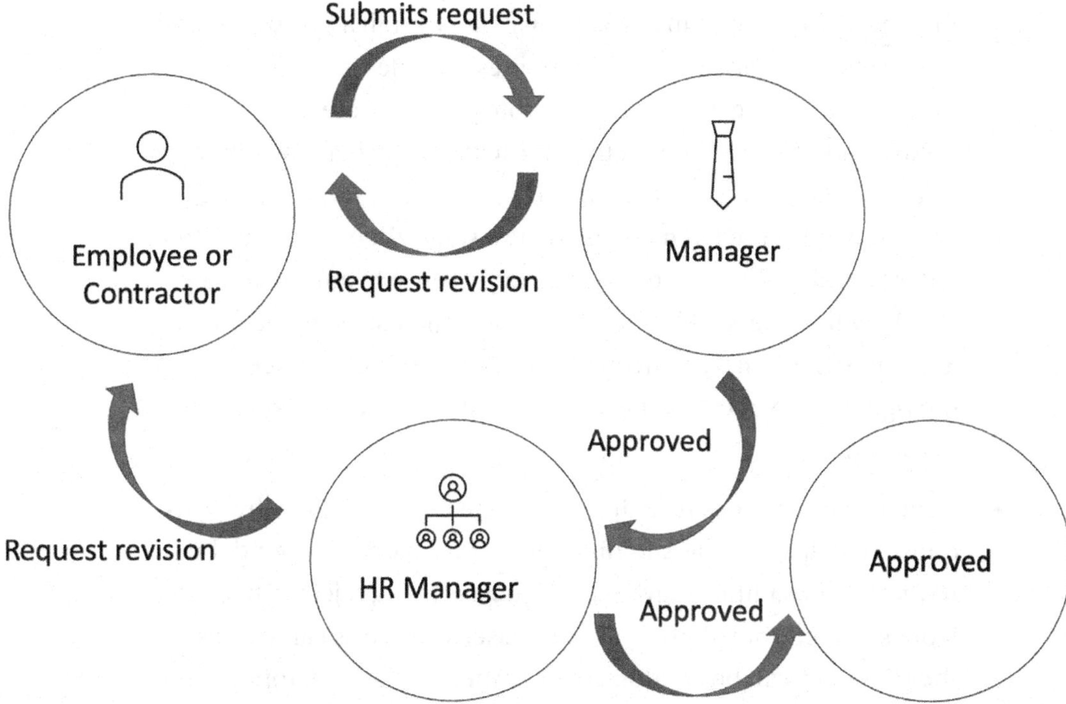


***Figure 3-1.*** *Vacation requisition workflow*

The preceding workflow generates logs for each vacation request that is used by the system as the request moves through the actors (employee, manager, and HR manager). Timestamps of each review stage are recorded along with the result of the review (approved/request revision). In process mining, the logs are loaded into the process mining tool. Most process mining tools have a visual interface that shows the movement of requests from one actor to another through time. Through this analysis, one can come to a better understanding of the process.

For example, in our case, you can find the answers to questions such as this with the help of the process mining tool:

- How many requests were rejected by the system due to the employee/contractor entering in more vacation time than allotted in the system? If many requests are getting rejected by the system, it can mean there is a lack of awareness for employees/contractors to know if they have enough time for a vacation. As a result, the workflow could be modified to display the vacation time before proceeding with the vacation request.
- Are any of the actors causing a delay in the overall approval process? If so, which ones? As you see, the request bubbles move about in the process mining visual, you can note places where the bubbles accumulate. This shows that that particular stage of the workflow is taking a long time. For example, if a lot of requests are getting accumulated at the HR manager review stage, then we can talk to the HR manager to see if it would make more sense to automate their checks within the workflow and eliminate their stage of the process altogether. This may ensure the requests are being actioned on at the right time. Nobody wants to wait for their vacation request to get approved.
- Which review stage is rejecting more than expected and why? This question helps in understanding the common issues around the request. For example, maybe the request is being rejected because it doesn't confirm HR policy. In this case, awareness campaigns of the HR policy can be conducted, or employees/contractors can be trained in HR employee training sessions.

Figure 3-2 shows the sample vacation requisition workflow with the vacation requests, as displayed within a process mining tool. Each individual bubble represents a vacation request. The HR manager actor has more bubbles than the manager actor, so more requests are being "stuck" at the HR manager stage.

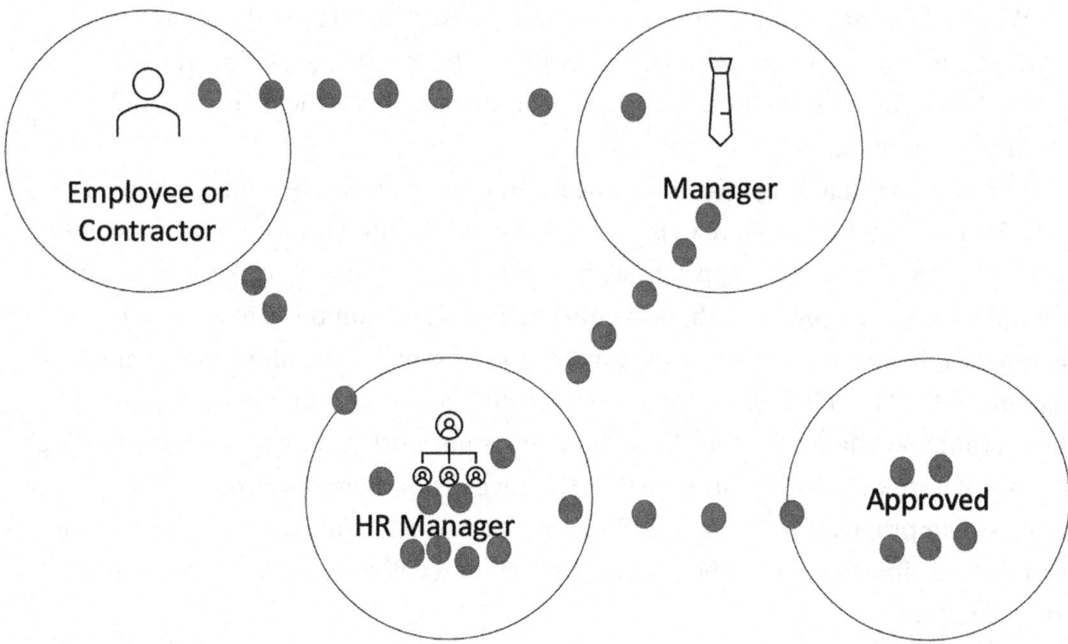


***Figure 3-2.** Vacation requisition process miner example*

As seen from the preceding example, insights can be derived with the help of the process mining tool that may not otherwise stand out in the data. From a risk management point of view, process mining can be used to understand high-risk activities within the organization and identify areas of improvement. Process mining can also be used as a way to confirm a finding and perform "what if?" or impact analysis.

## Continuous Auditing

Substantive testing, CAATs, and the process mining sections showed examples that used data to derive insights or report exceptions. These types of solutions most often help in conducting reactive analysis. The generated data is loaded into these solutions to derive insights, so the analysis is conducted after the bad event/exception has already happened. Although this is useful, there is a lag between when the issue occurs and when it will be "caught" by the tests. This lag can be expensive depending on the test and the process being tested. For instance, consider the case when an unintended payment to a vendor was caught over three months after it happened. The more time it takes to catch the exceptions, the harder it can be to correct the damage it caused.

What if there was a way to note exceptions and escalate them to the appropriate parties as they occur proactively? The lag between the event and the time the exception is "caught" by the tests would essentially be eliminated. This is the theory behind continuous auditing.

In continuous auditing, high-risk process areas or controls are monitored in real-time. So, when there is a failure or conditions are fulfilled for an exception, an exception is generated and sent to the appropriate personnel to spark them to take action. For example, consider a solution where terminated users are monitored to ensure they do not log into any of the company's resources after termination. Most organizations depend on the IT and HR processes to remove the access of terminated employees after their termination date. However, if this function is not working as expected due to a high volume of terminations or a break in the HR process, additional controls need to be put in place. Our proposed solution would run in real time and will grab the most current listing of terminated users in the system. The solution will monitor the access logs of all company systems.

A simple check would be to grab the terminated employees one by one and check the logs of the system to see if any terminated employee ID matches. If a match is found, it will send an exception to the Internal Audit's functional mailbox and email carbon copy (cc) to the company's security administrators. Internal auditors can then follow up with the security administrator to ensure that the unauthorized login is investigated.

## Conclusion

There are many solutions currently available in the audit space. Automation of testing and 100% data testing are methods that enhance the audit testing process. Process Mining helps to understand a process better and further understand the functioning of business processes. Lastly, fit-for-purpose technologies help address specific risks faced by the organization.

# CHAPTER 4

# Data Analytics

Analyzing raw data for insights has been widely used, especially since the dot-com bubble. **Data analytics** is the process of analyzing data to draw insights for decision-making purposes. Through data mining, raw data can be refined to get useful information. Although this concept was touched on before in previous chapters, specific data analytics methods were not discussed.

This chapter will explore the use of data mining in data analysis and data science. **Data mining** is a process of extracting and discovering patterns in large datasets involving methods at the intersection of machine learning, statistics, and database systems. A standard framework known as CRISP-DM has been around for many decades and has been effective in addressing the important aspects of data mining. We will look at this framework in detail.

For an auditor, it is important to understand the process of data mining when requesting data for audits and conducting testing. It is especially critical to use the cross-industry standard process for data mining (CRISP-DM) when applying data science and machine learning techniques in audits. Data needs to be retrieved, processed, and evaluated before it can be deployed in the analysis. Finally, data stories communicate the conclusions made about the data.

We will then look at some of the commonly used data analytics applications from an audit point of view. The four main types of data analytics applications explored in detail are descriptive, diagnostic, predictive, and prescriptive analytics. Finally, we will introduce the concept of data science and how it is different data analytics.

The main difference between data science and data analytics is storytelling. In data science, stories are used to focus on the insights and appropriate action can be suggested to the user to optimize decisions.



© Maris Sekar 2022
M. Sekar, *Machine Learning for Auditors*, https://doi.org/10.1007/978-1-4842-8051-5_4

# CRISP-DM

CRISP-DM stands for Cross Industry Standard Process for Data Mining and it is an open standard process model. It was introduced in 1996 to provide a standard set of guidelines for data mining projects. At a high-level, CRISP-DM consists of six main components, which are Business Understanding, Data Understanding, Data Preparation, Modeling, Evaluation, and Deployment.

The diagram in Figure 4-1 shows the CRISP-DM framework. CRISP-DM Process diagram by Kenneth Jensen (Own work).

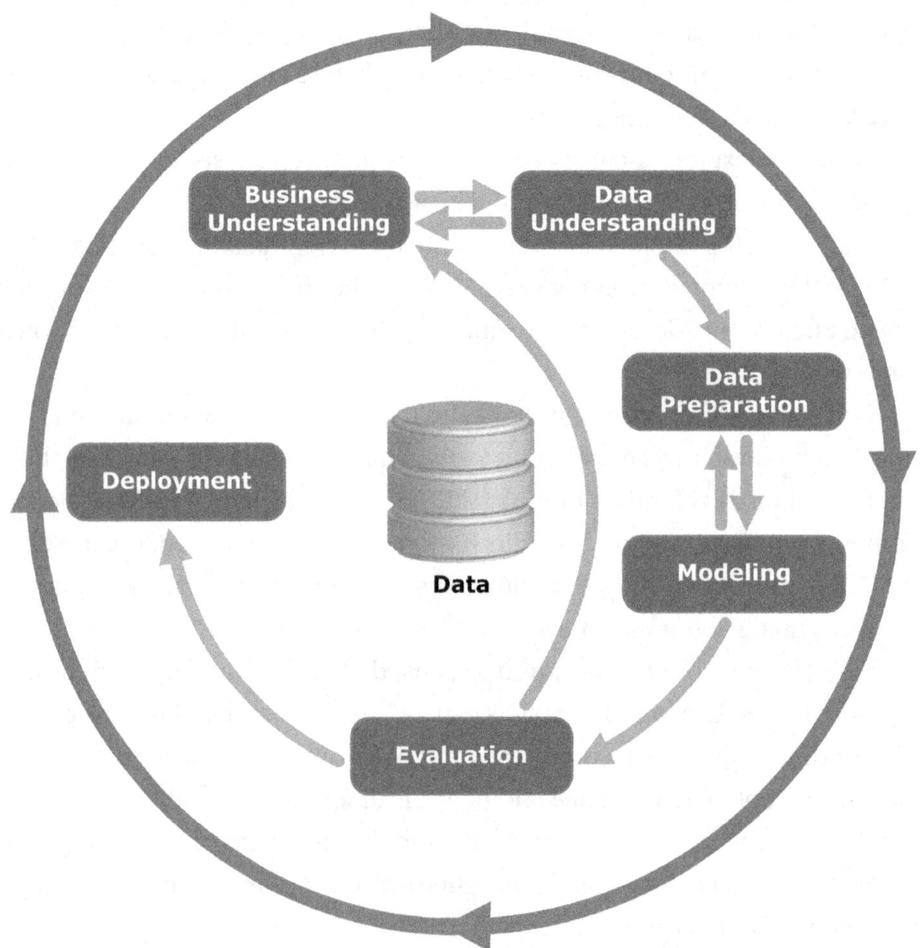


***Figure 4-1.*** *CRISP-DM Framework (CC BY-SA 3.0 [`http://creativecommons.org/licenses/by-sa/3.0`]), via Wikimedia Commons*

- **Business Understanding.** The business understanding component is the first step in a data mining project and it consists of developing an understanding of the business model. It is important to understand the objective of the data mining project. This component also clarifies the context and the business problem or question that needs to be addressed through data. For example, the business question could be "how many units of the product did we sell last quarter?" The objective of the data mining project could be to "find ways to help increase sales" or to "ensure raw materials are stocked to make the product." Understanding the business objective is critical because it has a holistic effect on the components that follow.
- **Data Understanding.** The data understanding component helps gain an understanding of the data itself in terms of checking it for completeness and accuracy. It is also used to explore the potential applications of the available data. The number of units of a product could be obtained in various ways, including from the invoice data, from the inventory data, or from the total sales and sale price of one unit. It may be better to obtain it in a particular way when compared to others based on the objective of the project. Through understanding the data, additional information about the business may come to light and be used to support the project.
- **Data Preparation.** The data preparation step is where the data is extracted, transformed, and loaded (ETL). As part of the data transformation, it may be cleaned for use in the next Modeling step. This step can take the majority of the time and often takes more than 50% of the total effort for the data mining project. Data cleaning transforms raw data into a format that can be used by algorithms.
- **Modeling/Analysis.** Once data is in a format that can be used by algorithms, it is ready to be modeled or analyzed. In this step, the data is modeled to derive insights about the data. The insights are used to predict or optimize the business objectives conceived during the business understanding section.

- **Evaluation.** As part of this component, results from the analyzed or modeled data are evaluated to ensure correctness. This can be done by comparing modeled results with expected results and calculating errors based on the type of algorithms used. For example, if the model is used to predict sales trend, historical known results can be compared to the predicted results to check the accuracy of the solution. After this step, it may be necessary to update the business understanding objective(s) and iterate over previous cycles (as shown in Figure 4-1) until accuracy is improved.
- **Deployment.** Finally, the optimized model or analysis is deployed to be used by end users. In the case of a dashboard, it is shared with authorized users so they can utilize insights in their decision-making process.

The six components are iterative in nature in that some of the steps can be repeated based on the requirements of the project and the quality of data. In addition, the iterative nature provides feedback mechanisms within the processes which improves the overall model.

# Data Analytics Audit Applications

The type of analytical techniques used can vary based on the question the user is trying to answer. Most data analytics projects can be divided into four main categories:

- **Descriptive analytics.** Descriptive analytics describes what happened in the past. It helps to explain the current state. The list of users who had access to a particular system is an example of this.
- **Diagnostic analytics.** Diagnostic analytics helps us to understand what is behind the data we are seeing. Its focus is more on answering the "why" question. For instance, the reason behind why certain users had administrative privileges can be explained with diagnostic analysis. This could mean support in the analysis to show the job function of the user. It makes more sense if the manager or the technical lead has administrative privileges when compared to regular users of the system.

- **Predictive analytics.** Predictive analytics helps explain what can be expected in the short term or long term given the current information. For instance, it can be used to predict the number of users most likely to be observed at a given time of the day. The prediction can be based on the historical system usage over the last year.
- **Prescriptive analytics.** In prescriptive analytics, recommendations can be made using data analytics based on the predicted values. For example, high risk modifications to the system can be recommended when the system is expected to be utilized by the least number of users.

Auditors generally use descriptive and diagnostic analytics the most to perform their analysis and complete their data verification. Predictive and prescriptive analytics is where machine learning and AI are normally employed because this involves data modeling based on historical data.

# Data Analytics vs. Data Science

Data science is generally seen as an evolution of data analytics. Yet there is one major differentiation between them. Apart from data science involving more statistical analysis, storytelling is the key differentiator. In data science, after the data is analyzed, it is shared in terms of stories that better connect with the end users of the analysis. For example, suppose you perform an analysis to look at the system usage of a high-risk application. In data analytics, the output would generally be a graph, perhaps a line plot, that shows the trend in user activity over a period in time.

In data science, we would look at the activity and derive stories from it, as shown in the following examples:

- Summer months show a decrease in user activity since many employees take a vacation during those months.
- Due to a one-week maintenance issue back in July of last year, we see a sharp decrease in user activity for July because users were unable to login to the system.
- Most users log in during regular business hours.

Data science focuses on deriving actionable insights that generate value for the business. They can also be used to help make key decisions to achieve business objectives.

# Conclusion

Data Analytics helps us to gain insights into the data through visualization and analysis. The CRISP-DM framework that was originally used to mine data in a structured manner can be leveraged for Data Analytics. Descriptive and diagnostic analytics are extensively used by auditors to help them verify data. The next progression in analysis is to tap into predictive and prescriptive analytics opportunities by implementing AI/ML audit applications. Data science builds on top of data analytics and uses results from data analysis to tell impactful stories.

# CHAPTER 5

# Analytics Structure and Environment

In a previous chapter (Audit Challenges), we discussed the importance of the adoption of data within all levels of the audit department. In order to adopt data, all internal audit members need to be trained to work with data. Hence, data literacy must be enabled in all levels of the team in order for it to succeed in powering the analytics needs for the audit department.

In this chapter, we will discuss some possible analytics organization structures that can be observed based on the overall organization structure. Data analysts and data scientists can either be embedded within squads or they may be part of functional units who are pulled into the business units based on projects' needs. No matter what organization structure is utilized, it needs to account for the existing audit department organization structure and incorporate the department leveraging the appropriate resources. We will look at how to do this in this chapter.

The organization environment needs to be conducive to the implementation of analytics within the internal audit department. In order for any analytics initiative to take shape, it needs support from other parts of the organization such as senior leadership. The tone at the top is a critical component to the success of analytics. There are some basic requirements from the organization spanning over technology, process, and people. We will explore some basic use cases in this chapter.

Senior leaders have an especially important role for encouraging and empowering their employees to use data analytics within their projects. Leaders need to define and measure the use of analytics within their projects or audits. We will look at some practical ways this can be employed within the internal audit organization.



© Maris Sekar 2022
M. Sekar, *Machine Learning for Auditors*, https://doi.org/10.1007/978-1-4842-8051-5_5

## Analytics Organization Structure

Implementing a new analytics structure means a sweeping change to existing structures and even processes. The type of structure and processes depend on the organization. So, we address these issues in the following sections.

There are many ways to organize the analytics resources. Choice of organization structure (or restructure) depends on the type of organization and the technical requirements of the audit department. Three of the common types of analytics organization structure are project based, functional, or business unit:

- **Project-based organization.** In the project-based organization, the analytics gaps of the audit team are met by utilizing a data analyst or data scientist from a pool of available data resources. The pool of data resources reports up to the analytics functional unit of the organization. Every audit consists of the audit team members, the data analyst or data scientist, and the manager who is accountable for the audit. The type of data resource needed will vary based on the needs of the audit. For example, if data needs to be visualized and analyzed for insights to drive business decision-making, a data analyst may be the right person for the job. But, if data needs to be analyzed for statistical inferences and stories need to be derived, a data scientist may be more suitable for the task. Data engineers may also be leveraged for more data-intensive big data applications, where heavier data processing is required. A data engineer prepares and maintains data for analytics or operational tasks.
- **Functional organization.** In the functional analytics organization type, some data scientists, data engineers, and data analysts may be assigned to work with the internal audit organization. This can be especially useful to train the data personnel in the domain knowledge of the organizational business function. For example, there may be a data analyst within the analytics business function of the organization that helps the supply chain team to build a vendor management dashboard. This data analyst can also work with the supply chain domain expert within the internal audit department for supply chain-related audits. A vendor management audit is a good example of where this can be employed. Even within the internal

audit organization, there may be domain experts for supply chain, accounting, IT auditing, etc. If the data resources are paired with the corresponding domain experts of the internal audit department, it leads to a better understanding of the objectives of the process being audited and further helps in audit testing.

- **Business-unit organization.** In this type of organization, the data analyst, data scientists, and data engineers are hired directly by the internal audit organization. A group of data resources may reside within the internal audit organization who help with the data analytics needs of internal audit. The data resources in this case may gain the domain knowledge of the audit department itself. For example, a data analyst hired within the audit department will be aware of the audit processes and data analytics needs of the internal audit team. The data analyst may not necessarily have the functional domain knowledge of the audits themselves.

Another common organization type which can be taken as the fourth organization type involves the use of a combined role – IT auditor/data analyst. In this type, IT auditors are hired with data analytics background or are trained in data analytics within the internal audit department. Since IT auditors have a technical IT background, they may be easily trained on basic data analytics techniques to support the internal audit organization. This type of organization may work for internal audit teams that do not leverage heavy data analysis, but it can get out of hand as the data analytics needs of the internal audit team expands. This is due to the dual role taken by the IT auditor.

# Organization Climate

For data analytics to be widely adopted and mandated in projects, the organization needs to recognize its importance and enable support within the organization. Apart from having specialized data analytics teams, it is important to support citizen developers. One way to promote data analytics projects is to communicate successful applications of data analytics to all parts of the organization. Another effective method to coach the organization is to utilize data analytics champions. Data analytics champions are people within the organization who are experts within their teams when it comes to using data to solve business problems. The champions need to be identified and guided by a data analytics Center of Excellence team (CoE). Team members can approach the data analytics champions assigned to their teams for any advice on their projects. Finally, lunch and learns can be provided by the data analytics champions or the data analytics CoE on specialized topics to train others on deriving insights from their data.

A common mistake with enabling adoption is to focus on small teams to implement data analytics initially with the plan of expanding into other parts of the organization over a period of time. The problem with this approach is that the support provided by the organization is not unified and, as a result, other parts of the organization may start using tools that are not standardized. Managing the data infrastructure in such a case may be a nightmare. The best way for a successful adoption of data analytics is to push standardized guidelines to the whole organization at the same time. The guidelines will state which tools are supported by the organization along with the recommended setup to retrieve information from the company's systems.

## The Role of Senior Leaders

Senior leaders play a pivotal role in ensuring data analytics is being utilized effectively by the team. When managing resources and designing a competency framework for the internal audit department, leaders should consider the use of data analytics. As per the International Professional Practices Framework (IPPF), the IIA has compiled the following standards:

- **IPPF 1210 – Proficiency.** "The use of competency frameworks is a successful internal audit practice that establishes a baseline of knowledge, skills, and experience for each level within internal audit. Included in an internal audit competency framework are the use of data analytics and technology in the internal audit process."
- **IPPF 2030 – Resource Management.** "Internal audit delivery and execution includes the defined methodology, IT risk and control concepts, and the use of data analytics and technology in the internal audit process to achieve greater audit coverage without the need to expand internal audit resource requirements. Alternatively, internal audit will be required to actively seek to obtain or train staff with skills and focus on data analytics to support internal audit activities."

## Conclusion

In order for analytics to succeed in an organization, it needs to be part of an environment that favors analytics growth. The organization structure and data analytics support (in addition to data maturity) are two factors to consider as part of the organization environment. Senior leaders need to consider the data analytics needs of their audits and ensure appropriate resources (people, process, and technology) are in place to support their teams.

# PART II

# Understanding Artificial Intelligence

CHAPTER 6

# Introduction to AI, Data Science, and Machine Learning

This chapter will dive into the world of artificial intelligence, machine learning, and data science. It will talk about the relationship between these components. When we talk about AI, it is generally defined as intelligent systems that mimic human intelligence. There is a certain element of autonomous behavior to the systems in that they tend to not need any manual intervention once the systems are implemented. The AI systems may *learn* ways to perform certain actions based on historical behavior. Machine learning algorithms help the AI system to learn from past data that are collected using data acquisition systems or sensors connected to the AI system.

## A Self-Driving Car

Consider a self-driving vehicle as an example. The self-driving system architecture can consist of infrastructure, vehicle sensors/cameras, vehicle motors, central control unit, satellite communication, and cloud backend.

The road signs, markers, and other specialized roads that are needed by the intelligent system are covered by the infrastructure. These may need to be modified to a certain extent for the sensors or cameras to capture the road conditions efficiently in a particular environment or road condition. Cameras may not capture vital images in low sunlight settings. Image captures may be used to read traffic lights, pedestrian signs, road markers to align the car on the road, etc. There may be a need to integrate traffic lights with self-driving cars in order to improve the decisions made by the car. Driver-assist features such as maintaining distance and lane assistant to keep the car from departing



© Maris Sekar 2022
M. Sekar, *Machine Learning for Auditors*, https://doi.org/10.1007/978-1-4842-8051-5_6

the lanes unexpectedly use lasers to guide its actions. During rain, snowfall, or fog conditions, these lasers may not work as expected. Cheap radio frequency identification (RFID) tags can be embedded within roads to help support the lasers and can provide a more accurate state of the road.

Specialized sensors, cameras, and motors may need to be equipped in the car to collect data from the environment to make decisions based on the real time data being collected. Laser sensors to read road markers, cameras to read traffic lights, and motors in the car to control the steering wheel are some example use cases of this equipment. One important point to note is even though data is collected for a particular purpose it may need to be integrated with other sensors for building robust and safe systems. For instance, if the road markers are obstructed due to snow on the road, the vehicle's intelligence systems may need to rely on the cameras to ensure it is not departing lanes. In addition, real time data has many challenges. One major challenge with real time sensors is it is often missing values due to sensor issues or other unexpected data capture problems. This must be kept in account by the intelligent system when making decisions. The steering wheel motors may need to be calibrated regularly to ensure optimal usage. If the cameras, sensors, or motors are not functioning as expected, they need to be notified by the vehicle user in a safe and reliable manner.

The central control unit is the main decision system of self-driving car. This unit processes the data captured through the sensors and cameras to make data-driven decisions. They also help control the motors based on decisions to control the navigation system, brakes, acceleration, etc. This unit may use machine learning algorithms to *learn* from the environment and make effective split-second decisions on behalf of the driver. The central control unit may identify obstacles, objects, including humans, and other vehicles using computer vision machine learning applications. It can also use machine learning to adapt speed based on the environment such as based on the objects on the road or inclination of the road.

Another important job of the central control unit will be to acquire data and store it as efficiently as possible so that it can take appropriate decisions based on the collected data. This may include collecting and sending data to cloud infrastructure to analyze data. In some cases, machine learning tasks that need extra processing capabilities may be sent to the cloud to process using faster processors for lower priority functionalities.

Cloud backend servers are mainly used as a storage mechanism for the data collected using the vehicle intelligent systems. The data collected may be useful for analysis at a later time. Cloud servers may also be used as a processing system for tasks that may need additional processing capabilities.

The vehicle may rely on satellites for utilizing maps using Global Positioning System (GPS). The map routes, speed limits, and navigation system can be integrated with the vehicle intelligent systems in order to assist it in the decision-making process. Posted speed limits by the Regional Transport Authority (RTA) can serve as an additional cross check tool when cameras are not able to read road signs during poor weather conditions.

Data science can be used in vehicle intelligent systems before machine learning algorithms are applied. For instance, Principal Component Analysis and Data Scrubbing (or Data Munging) may take place in order to get the data into tables that capture one observation per line and simplify the data being captured. This is important for machine learning algorithms to be able to work on the dataset. Data Munging involves the cleaning up of raw data and the application of data transformations so only the required data is extracted to work with.

Figure 6-1 shows an example of Principal Component Analysis (PCA) that can be used in vehicle intelligent systems. PCA is a dimensionality reduction algorithm that transforms a large set of features into a smaller one that still contains most of the information.

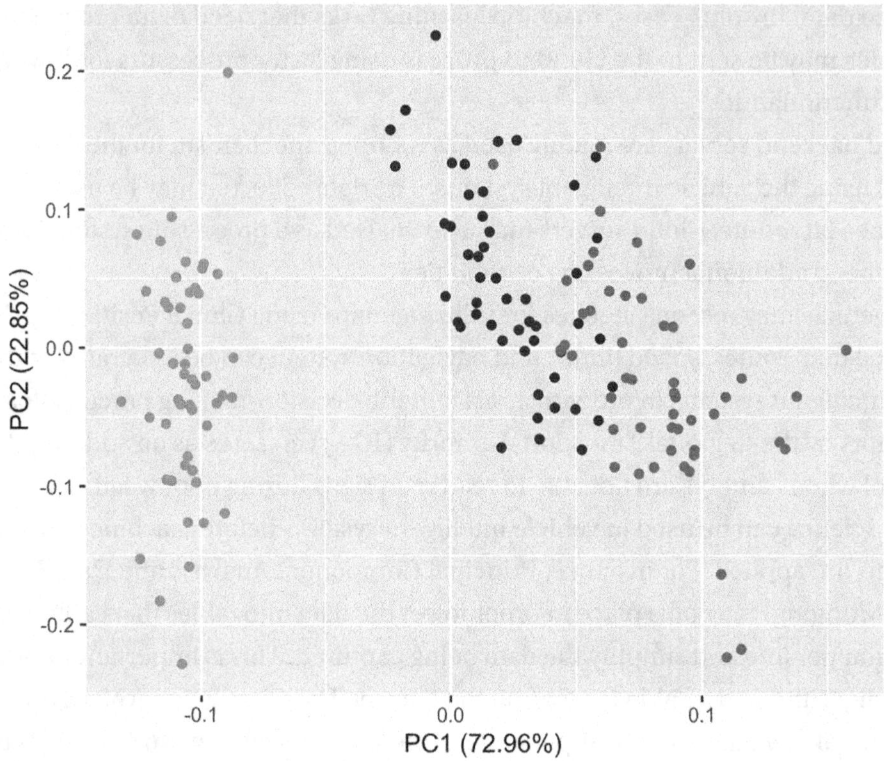


***Figure 6-1.*** *Example Principal Component Analysis*

# Components of an AI System

An artificial intelligence system consists of two components, namely, machine learning (ML) and Robotics Process Automation. ML embraces two distinct techniques, namely, unsupervised learning and supervised learning. Unsupervised learning occurs without guidance. Supervised learning learns from the data using labeled data. Unsupervised learning is largely composed of clustering ML techniques. Supervised learning involves ML algorithms such as classification, regression, and deep learning. Robotics Process Automation are normally carried out by specialized tools or scripting languages.

Data science is utilized to derive insights using scientific methods, systems, and algorithms. Data science can be used by itself or it can also be utilized in conjunction with ML techniques. Data science can be further divided into Data Analytics and at the lowest level, Data Mining. The components will be examined further in the following to understand their functions.

The diagram depicted in Figure 6-2 shows the relationship between artificial intelligence, ML, and data science. It also shows other smaller components that are part of ML and data science such as Deep Learning, Data Analytics, and Data Mining. One other component of AI, Robotics Process Automation, is also shown in the diagram.

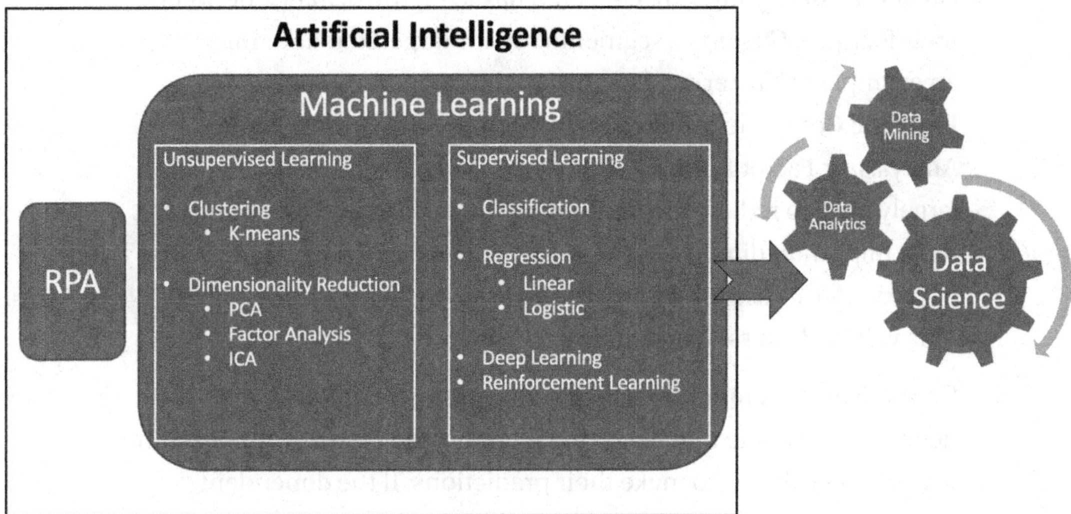


***Figure 6-2.** Components of AI*

- **Artificial Intelligence.** Autonomous machines that mimic human intelligence are broadly defined as artificial intelligence. A self-driving car system is an example application of AI. An automated invoice processor is another use case of AI.
- **Machine Learning.** Machine learning supports AI intelligent systems by using historical data to make decisions or optimize a process. These are divided into supervised and unsupervised learning, which is delved into deeper in the following.
- **Robotics Process Automation.** It is the other method broadly defined under AI. It is the utilization of specialized tools and scripts to automate manual tasks. For instance, an accounts payable mailbox can be monitored for invoices. Whenever an email with an invoice is received, it can read the invoice and populate the system tables. The information will be reviewed by the accounts payable staff before it is set up for payment. The invoices may need to be sent in a predetermined format by the sender in order for the translation to occur properly.

- **Unsupervised Learning.** A class of machine learning techniques that deal with unlabeled data. Unsupervised learning is used to derive hidden insights from data. Clustering and Dimensionality Reduction are two of the most common types of unsupervised learning. In clustering, data points are clustered based on the distance between their features. Customer segmentation for targeted marketing is an example of clustering. Dimensionality reduction can involve Principal Component Analysis (PCA), Individual Component Analysis, or Factor Analysis techniques. These techniques are largely used to reduce the number of dimensions and focus on the most important dimensions to work with. Anomaly detection is an important use case for unsupervised learning in the audit domain. This will be discussed in detail later.
- **Supervised Learning.** In supervised learning techniques, a dependent variable that needs to be predicted is used by the machine learning algorithms to make their predictions. If the dependent or target variable is a categorical variable (Yes/No, Male/Female, etc.), classification algorithms can be used. Predicting the gender based on the height and weight is a sample use case of classification algorithms. If the variable to be predicted is a quantified value that changes based on the inputs, regression algorithms will be utilized. Predicting the stock price based on the historical stock prices is an example application of regression techniques. Deep learning is a subset of a broader family of machine learning algorithms based on artificial neural networks with representation learning. Deep Learning can be supervised, semi-supervised, or unsupervised.
- **Data Science.** Data science is the application of methods, systems, and algorithms to derive insights from data. It is an evolution of Data Analysis. It includes statistical methods and sharing of impactful stories that are supported by data. For instance, consider a particular workstation on the network is being targeted regularly by hackers because it has not been patched in a while. We can derive the probability that it will be attacked again in the future. Say if 80% of the attacks on the network go to this workstation – tougher

restrictions can be put in place to mitigate the risk in the future, or the workstation can simply be patched with the latest update. We can also answer questions such as "Is it worth my money to invest in upgrading my workstation?" and "Will the attacks keep increasing based on the current trending?"

- **Data Analysis.** Data Analysis is composed of techniques used to clean, transform, and analyze data. It can be used to support a specific business objective that has been pre-determined prior to the commencement of the data analysis. It can help answer questions such as "How many workstations have been patched?" and "How did I invest my budget for the current year?"
- **Data Mining.** At the heart of data analytics/data analysis, data needs to be extracted from data sources. They need to be retrieved and processed to be able to perform data analysis techniques. Data mining is the process that helps extract data based on the analysis being performed. For example, for an expense management analysis, we would need to connect to the expense management system (which may be embedded within an ERP) to retrieve all expenses for the time period being analyzed. Once the data is retrieved, it may be loaded into a data warehouse and connected to a data analysis tool. The expenses can then be analyzed using a data analysis tool.

# CRISP-DM for Data Science

CRISP-DM stands for Cross Industry Standard Process for Data Mining. We introduced it in an earlier chapter as a general framework that can be used as a guideline for Data Mining projects. In this section, we will look at how CRISP-DM can be applied to data science. CRISP-DM's emphasis on the iterative flow of the steps involved ensure it is the right candidate for data science–related projects. Data science is highly iterative by nature when compared to data analytics. For instance, data science phases such as Exploratory Data Analysis and Model Evaluation often end up leading to previous phases rather than linearly to the next phase. Many data science teams have successfully used the CRISP-DM framework in conjunction with agile methodology to develop effective data science projects.

The diagram in Figure 6-3 shows the CRISP-DM framework for data science projects.

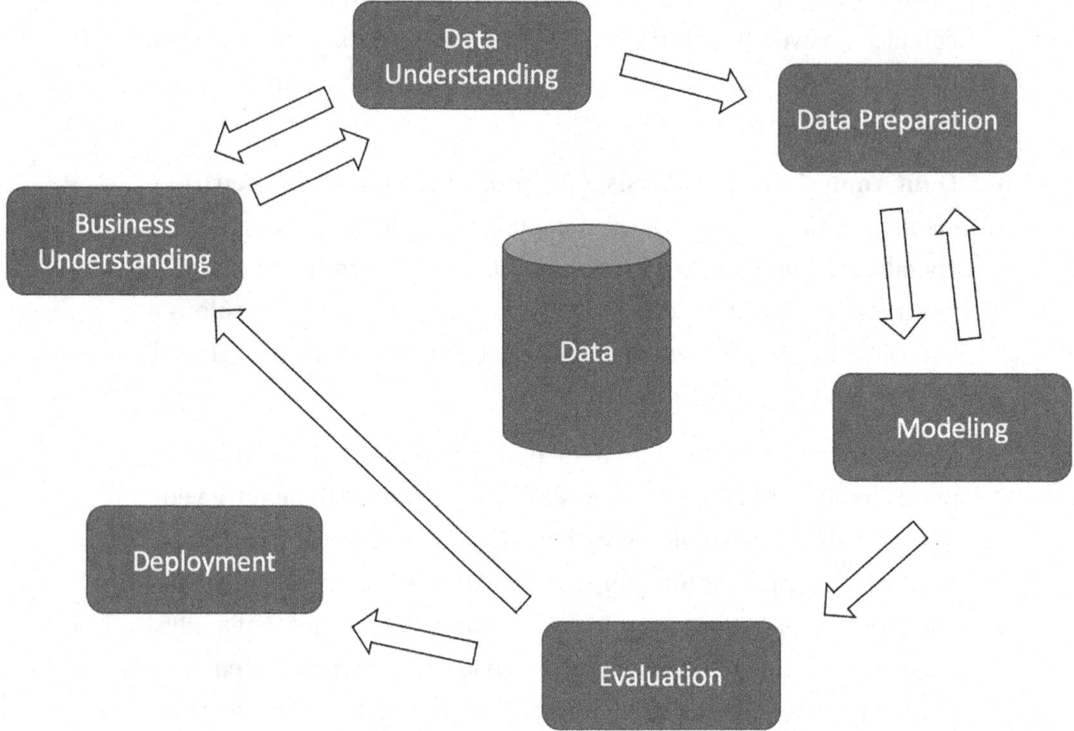


***Figure 6-3.*** *CRISP-DM framework for data science*

- **Business Understanding.** This step helps in clarifying the business objectives and understanding the business problem that the data science project is trying to solve. This concept is very similar to one used in the regular CRISP-DM framework application.
- **Data Understanding.** Data must be retrieved and analyzed to see if it can help solve the business problem at hand. In this step, the data science team collects the data and probes it to get a clearer understanding of the problem. Exploratory Data Analysis (EDA) is performed to get a deeper insight into the process under question. As a picture is formed with the supporting data, the business problem may need to be revisited in the previous step. Suppose the original business question was to find the high-risk areas for fraud in a vendor payments process. We need to first define what *high risk areas* mean

based on the data collected. Now suppose we have retrieved the data containing payment amount and payment date. One high risk area based on these features would be if payment amount is high and the payment date is not at the end of the normal pay period. After we have explored the payment amount and pay periods for all provided payments, we may find that many payments fall in this criterion. In this case, we can go back to the business understanding step of CRISP-DM and redefine the business objective. Instead of saying *high risk areas,* we can define what makes a high-risk area in terms of the features that are available. Data quality is also inspected in this step. The collected data is checked for features that are empty and those that are in need of a data transformation in order to be useful for the project.

- **Data Preparation.** In this step, a data science project team performs *data wrangling* based on the data quality and insights gained from the previous step. Data may need to be filtered down to only the features required and may involve identifying and removing redundant columns or reducing the complexity of the features. For instance, dates with timestamps may not be necessary. Instead, a short date format that displays only the date might be enough for the job at hand. New fields may need to be derived from existing fields. Suppose we need to convert a score of 0–10 to a Red, Amber, and Green (RAG) scoring for easy interpretation. We can define 0-3 as *Red,* 4-6 as *Amber* and 7-10 as *Green.* We may need to consolidate columns from multiple tables in order to make it easy for analysis. For example, if only the employee ID is defined in Table A and Table B contains the employee's first and last name, we will need to join Table A to Table B to create a consolidated Table C that has the employee ID as well as their corresponding first and last names.

- **Modeling.** This step involves the building and assessment of models based on the applicable modeling techniques. The type of algorithm to use, such as k-means, linear or logistic regression, artificial neural network, etc., will be chosen based on the type of analysis performed. For instance, take clustering using the k-means algorithm.

Parameters for the algorithms may need to be optimized in this step. K-means requires a value for k which is normally between 1 and 10. There are multiple ways to come up with the optimal k value to use. The input datasets will need to be split for supervised machine learning algorithms. In this case, the dataset will be divided into training, test, and validation sets. The training set will be used to train the model. The test set is used for generalizability because the model (algorithm) has never seen the dataset before. The validation set is used to squeeze out extra performance by tweaking the trained model.

- **Evaluation.** In this step, the data science project's arrived model from the previous step is evaluated to ensure it meets the business objectives set forth at the beginning. The supporting results are analyzed to ensure they support the appropriate reasons for choosing the final model and they are captured in a summarized report. At this step, we may choose to optimize the model even further by iterating with other parameters. When there is no more optimization to be done, we can proceed to deploy the model. The test dataset can be used to gauge the model's performance in the *real world* setting.
- **Deployment.** In this step, the optimized model is deployed for utilization in the data science project. After deployment, we may find additional details for features that may not have existed in the dataset provided. In this case, we would iterate over the framework from step 2 (Data Understanding) to ensure the new details are captured and the other steps are reperformed to ensure they account for the new data. The model's performance needs to be continuously checked for post deployment for ongoing projects to ensure its variability over time is captured and accounted for. In certain scenarios, if the data structure is altered over time – a new model needs to be retrained to ensure its accuracy is preserved.

## Domain Knowledge

In the previous section, the modeling step of the CRISP-DM framework requires input from domain experts in order for the model to be effective. The model has to cover the edge cases that domain experts may be aware of but may not be reflected accurately within the data. An edge case is a problem or case that only occurs at an extreme operating parameter, for instance, maximum or minimum. Domain experts are people who have been trained in an area. The knowledge gathered could be from formal educational training, years of work experience, and/or company-specific information.

For an artificial intelligence system to be effective, the system needs to take into account the functional knowledge of the process in question. For instance, suppose we design an AI system to perform predictive maintenance on fertilizer factory equipment. Data from the equipment, including machine running times, offline times, and turbine speed, may be gathered. Data about the equipment such as when the machine was purchased, warranty information, and maintenance information may also be captured. But the common pitfalls, possible cause and effects, and black swan events specific to a location may be hard to see within data.

For example, the machines may be seasonally utilized, which can be confirmed from the data gathered by the machines. But why it is seasonally utilized can only be known by a domain expert. The machines may be seasonally used because it may depend on the demand received from the company's customers, the farmers. The demand for fertilizers may have gone up at a specified time (say six months) before harvest. If the farmers expected ideal weather that has not been seen in decades, they may want to plant more crops, driving up the demand. In this case, the fertilizer factory equipment may need to be utilized more than their average use in the past and this could mean more frequent maintenance work. If the fertilizer company is able to predict this, they may be able to minimize their equipment downtime and maximize their revenue. Data scientists have to learn to communicate effectively with a non-technical audience (domain experts) in order to help them address the most important business objectives. Domain experts, on the other hand, need to be able to leverage data scientists' technical expertise to help them answer business questions that will bring the most value to the organization.

The diagram in Figure 6-4 shows the different quadrants and their properties based on the level of technical expertise and domain expertise available to work on a project. The lower the data science and domain knowledge the more primitive the data analysis becomes (represented by the Primitive data analysis quadrants in the Figure). If the data science expertise and domain knowledge are utilized, it leads to more predictive analytics (Predict modeling quadrant). Lastly, if both domain knowledge and data science expertise are not used, the data rests with no value (Untapped Data accumulation quadrant).

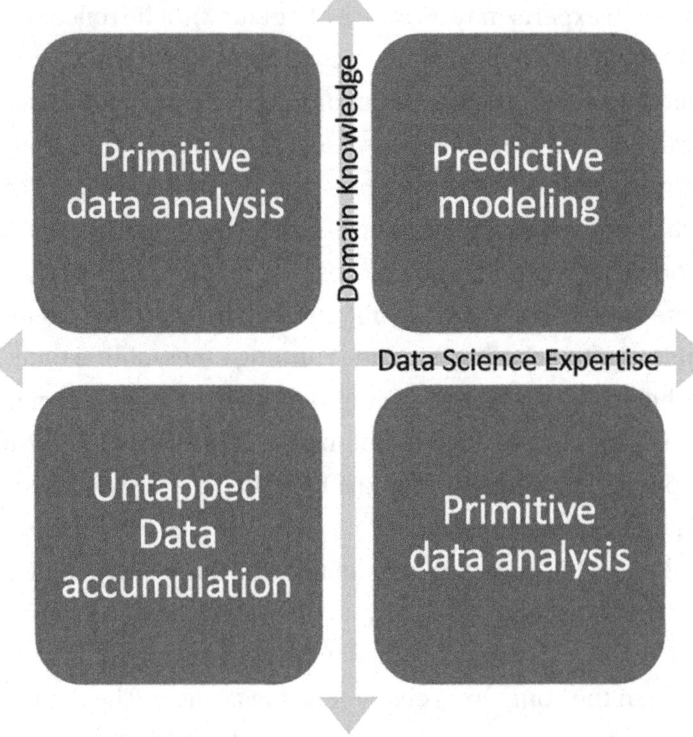


***Figure 6-4.*** *Domain knowledge vs. data science expertise*

# Payment Fraud/Anomaly Detection

In this section, we will talk about an example use case for applying an unsupervised machine learning, specifically k-means clustering, to the field of anomaly detection. The anomaly detection system, when applied to the area of payment analysis, can be an effective payment fraud prevention system. Fraud is defined as the wrongful or criminal deception intended to result in financial or personal gain. It can be hard to detect payment fraud using basic filtering techniques and search criteria. Since the fraudsters are aware of these filtering techniques, they try to find more elaborate ways to commit fraud.

According to the *2018 Global Study on Occupational Fraud and Abuse - Report to the Nations* by the Association of Certified Fraud Examiners (ACFE):

- Fraud resulted in over seven billion dollars in total losses worldwide.
- Financial statement fraud schemes were the least common and most costly, resulting in 10% of the total cases analyzed.
- Data Monitoring/Analysis were correlated with the largest reductions in fraud loss and duration - 52% lower losses and 58% faster detection. Yet, only 37% of victim organizations implemented these controls.

These surprising statistics are good reasons to apply machine learning to detect fraudulent payments.

Readily available vendor payment data such as invoice date, invoice number, vendor number, payment date, and payment amount can be used as data to design an anomaly detection system. After extracting these details from the company's databases, they will need to be transformed so that one payment is captured per line along with their corresponding features. This is crucial for machine learning algorithms so they can process the information.

The features (invoice date, invoice number, etc.) will need to be created to ensure a number is captured to represent each detail. For example, to convert the invoice date and payment date into a feature, we can find the difference between payment date and the invoice date. Since the payment always occurs after an invoice is received by the company, we will get a positive value to represent the number of days it takes for the company to pay a vendor. The vendor number can be used as is or it can be used to assign a score based on other vendor information such as location, size of the company, or purchase order amount. For instance, a vendor of a smaller size may be at a higher risk of going out of business and delaying the organization's projects if projects depend on these types of vendors to do business.

It is important that the input file going into the ML algorithm captures exactly one vendor per line along with the corresponding features. If multiple lines of the same vendor contain other feature values, the algorithm will not be able to generalize the dataset. After the features are extracted and each line in the input file contains information about exactly one vendor, we can proceed to the next step of creating a machine learning model. For our objective of detecting anomalous payments, we would like to cluster payments that look similar so that payments that are dissimilar to existing payment behavior are easily identified by the algorithm and can then be investigated.

Clustering will enable us to do just this. We can use k-means clustering using libraries that are provided by popular data science programming languages such as R and Python. For example, in Python, the following two-line code snippet is all we need to create and train a k-means clustering object using the scikit-learn library. The first line creates the model and the second line fits the model by training on "X":

```
kmeans = KMeans(n_clusters=2)
kmeans.fit(X)
```

The preceding code creates two clusters. We can define as many clusters as we would like but there are methods to determine the optimal number of clusters. For example, we can use the *elbow method* to calculate the clustering score for up to ten clusters to find the optimal number of clusters that provide the best clustering score. After applying k-means clustering, we hope to clearly see payments under each cluster through color coding and/or different markers. It may not always be clear to distinguish the clusters. If clustering works as expected, we will be able to attribute each payment to exactly one cluster. After this, we can look for outliers or points that are far away from the clusters and analyze those payments further.

The diagram in Figure 6-5 shows an example plot after applying k-means clustering and assigning each observation to a cluster. Each of the clusters are highlighted with a circle and can be distinguished by the shape of the data points. There are a total of three clusters in the diagram:

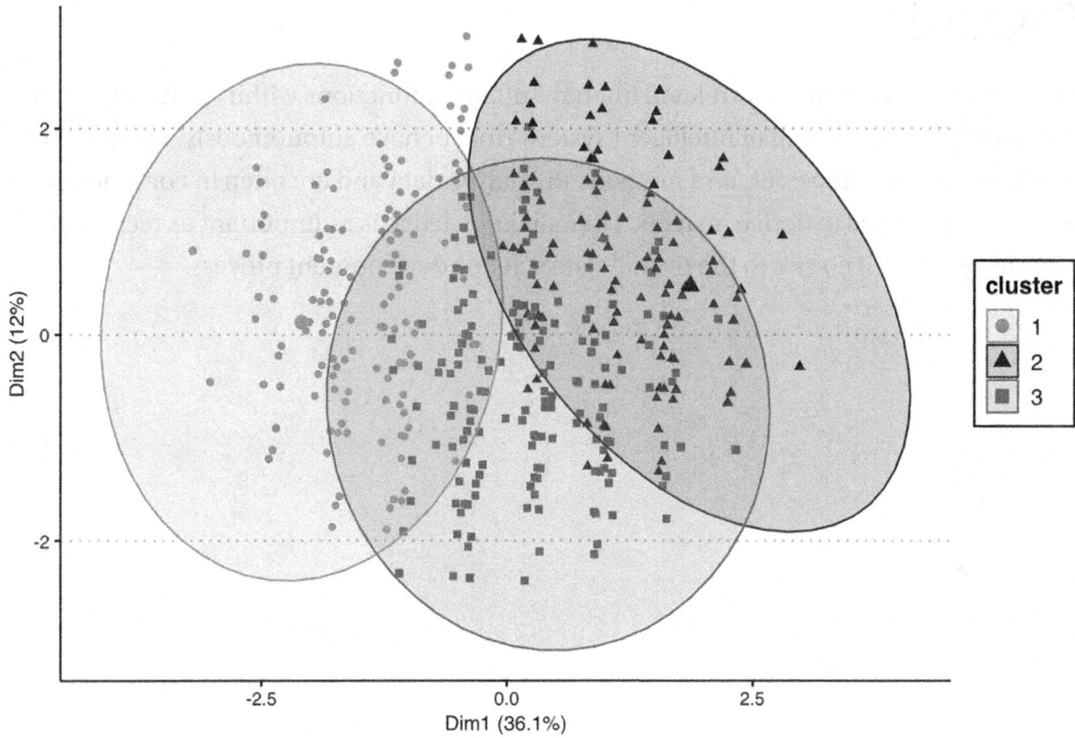


***Figure 6-5.*** *K-means clustering example*

For any machine learning application, it is important to note that although some models can provide a high level of accuracy and precision, there are chances of finding false positives. False positives are those that are predicted to be correct but in actuality they are false. So, it is always crucial to ensure the findings or exceptions are investigated using the appropriate channels in order to confirm if the findings are indeed findings. This can be a simple touch point with the business owner who is most in tune with the process. For instance, for our payment anomaly detection system, we can confirm the findings with the accounting team to ensure the potential fraudulent payments are correctly identified. If the accounting team reviews the payments and finds these are not fraudulent payments the machine learning model can be tweaked to ensure this particular case of false positive is not identified in future. It is important to note that the optimization of the machine learning model development process is a highly iterative process.

## Conclusion

ML and data science are interrelated but have different functions within an AI system. AI is the overarching domain of intelligent systems that behave autonomously. Data science provides systems, processes, and methods to analyze data and are often in conjunction with ML algorithms to derive insights. Domain knowledge is as important as technical knowledge when it comes to the overall data science development process.

# CHAPTER 7

# Myths and Misconceptions

In this chapter, some of the most important myths and misconceptions will be discussed. This chapter intends to break down the barriers to the practical application of data science techniques in our everyday work environment. Unfortunately, in certain situations, terms like artificial intelligence, machine learning, and data science have become household names and buzzwords that are used more as conversation starters.

On one side of the spectrum, some people have a lack of awareness of the subject. On the other side, some people are quick to dismiss that the field of AI is a new fad that will dissolve as quickly as it came. As proven by many successful companies, AI is here to stay and is a rapidly developing industry. Billions upon billions of dollars are being invested in creating this area actively. And this trend is sure to continue as countless practical AI applications are discovered over the next decade. If utilized wisely, the potential AI applications in the areas of risk management and auditing are immense. As is the case in almost all domain areas, risk management and auditing pose their unique challenges. We discussed many of these challenges in an earlier chapter.

Along with domain challenges, there are also challenges associated with bringing awareness to the subject. When building a business case to implement an AI or a machine learning solution, it is essential to address some commonly asked questions. If they are not handled properly, the project may come to a halt quickly.

The myths discussed, along with the facts discussed in this chapter, can be used to answer some of the commonly asked questions to build a better business case when implementing a new AI/ML learning system.



© Maris Sekar 2022
M. Sekar, *Machine Learning for Auditors*, https://doi.org/10.1007/978-1-4842-8051-5_7

## Myth #1: You Need an Advanced Degree to Be a Data Scientist

When implementing data science projects, a question often asked is: Do I need a data scientist or can I train someone on the team to perform data science tasks? The answer to the question depends on the scope and nature of the data science work.

In the field of auditing, some form of data analytics work may already be supported by some team members. Data science can be seen as a more advanced form of data analytics work supported with statistical analysis and storytelling. If data analytics experts are already on the team, it may be better to train the existing data analytics staff on data science concepts. The domain knowledge that the data analytics staff have gained from the team can be seamlessly applied for data science projects. Practical data science courses are available in all formats, including online, blended, classroom, and one-on-one. Mastery of such courses may be enough for practical application on "quick win" projects.

An advanced degree such as a PhD in a quantum field may not be required in most cases. Some projects, however, may require a good grasp of feature engineering, parameter tuning, and other advanced concepts that involve a deeper understanding of data science. In these cases, advanced knowledge of statistics or mathematics may be beneficial.

There has been a big push by many analytics providers to make data analytics more accessible to all. As part of data democratization initiatives, many analytics tools market themselves as a no-code or low-code solution. They are also making it easy to embed data science into existing tools and within visualizations. This overall push is against the fact that you need an advanced degree in order to derive insights from data.

## Myth #2: Correlation Implies Causation

If there is a correlation between two variables, it does not necessarily mean causation between them. For example, suppose the two variables are the height and weight measurements of students in different classrooms. When we look at the measures of one classroom, we may find that the weight increases whenever the height increases in general. From this finding alone, we cannot conclude that the weight is caused by height. However, we can say there is a high degree of correlation between height and weight. In another class, you may find that there are a lot more tall and skinny students. So, the trend, in this case, may be that as the height increases, weight may decrease or stay the same. Height does not cause the weight to increase. Eating food causes the weight to increase. More food being consumed implies that weight will increase.

It is essential to note the difference between correlation and causation. Correlation is a statistical technique that helps us understand how closely two variables are related and change together. Correlation just implies there is some relationship between the variables; nothing more and nothing less. On the other hand, causation tells us that if we change one variable, it will cause another variable to change. Some variables may look like one independent variable causes the other dependent variable. Still, there may be a third variable influencing both of them and it may be the real explanation behind the cause of both the independent and dependent variables. This third variable is called the confounding variable. For instance, it may be observed that whenever someone applies sunscreen, they may be wearing shorts. The application of sunscreen does not cause you to wear shorts. It was the hot temperatures that caused you to apply sunscreen and wear shorts. In this case, the sunscreen is the independent variable, and the wearing of shorts is the dependent variable. The outside temperature represents the confounding variable.

Figure 7-1 shows the relationship between the independent, dependent, and confounding variables.

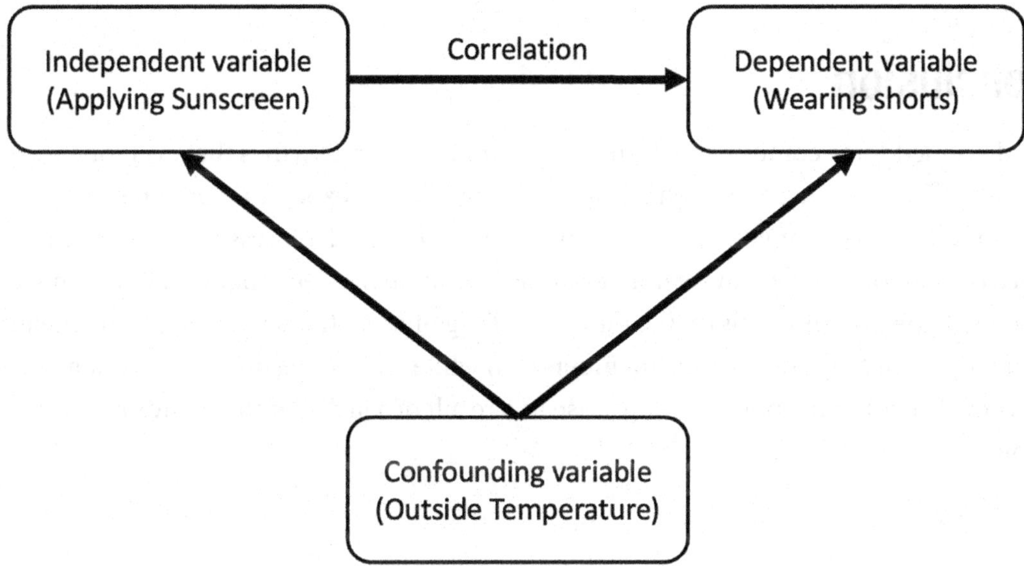


***Figure 7-1.*** *Relationship between variables*

## Myth #3: The Model Building Is the Most Critical Step

Although building models is part of a data science or machine learning project, this step is **not** the most important one. There is a lot of emphasis on building models and tweaking them to predict or derive hidden insights from data. This is true to a certain extent, but among all the steps involved in a data science project, the most important one would be to scrub the data and ensure it is without data bias. Inaccurate data being fed to the modeling step means the output will be useless – *garbage in, garbage out.* So, the most important thing is to ensure the data being fed in is free from all forms of known bias and clean to the point where it is easy to be consumed by the machine learning algorithm.

It is not uncommon to spend over 50% to 80% of the total time of a machine learning or data science project in the cleaning step (or data preparation) step. Model building is an iterative process and may be a little time-consuming when looking at the total number of iterations needed to come to the final model. But model building still takes less time when compared to transforming the data from raw unprocessed data to data that can be easily consumable by the model building step.

## Conclusion

We discussed three common myths of data science. The first myth is that everyone who practices data science needs a PhD degree or should be an expert in mathematics or statistics. The second myth has to do with cause and effect. It addresses the question: Does correlation between two variables imply causation between them? The third myth dealt with the overemphasis on the model building phase. These myths are by no means exhaustive, but they are essential to understand while embarking on the data science journey. The misconceptions mainly arise as a result of a lack of understanding of the subject.

# CHAPTER 8

# Trust, but Verify

In this chapter, we will look at ways of verifying findings from audit testing. Particularly, audit findings that involve machine learning and data science projects will be discussed in detail. The concept of "Trust, but Verify" will be introduced first. This will be followed by an explanation of why it is important to verify findings, including some of the relevant challenges of machine learning models when it comes to reperformance of testing. The chapter will end with an overview of the various sections used to report findings for data science and audit projects. The sections of the data science project will be integrated with the audit projects to ensure it supports the audit verification process.

## What Is Trust, but Verify?

In simple terms, "Trust, but verify" stresses the importance of verifying a claim through factual information rather than solely depending on the claim itself. It is based on a Russian proverb that has been famously used by many United States Presidents. President Ronald Reagan used "Trust, but verify" in 1986 to handle the arms-control negotiations with Russia. Since then, his successors have used the same concept to convey a similar message. At the heart of the principle, a hypothesis is being verified by looking for evidence that supports it. Once evidence is obtained, the hypothesis becomes a fact that cannot be disputed by others.

A demonstration of the principle will make the concept clear to understand. Consider an Information Technology (IT) audit on the organization's IT vendor management process. The IT auditor assigned to the audit has performed testing based on his audit program. An audit program contains the tests and procedures the auditor performs to test the vendor management process. As a result of the testing, the IT auditor has found some gaps in the vendor management process. Specifically, the auditor has identified a couple of vendors that could have been potentially paid twice due to an incorrect vendor setup. In this case, even though the evidence seen from



© Maris Sekar 2022
M. Sekar, *Machine Learning for Auditors*, https://doi.org/10.1007/978-1-4842-8051-5_8

the data strongly suggests that the identified vendors were paid twice, it is yet to be confirmed by the business that this is indeed the case. Data by itself can be misleading. In this example, upon probing further, it was found that the vendor was paid just once. The vendor changed their address and this resulted in a new vendor to be set up in the system. An "Active" flag in the data identifies vendors that are inactive in the system. By checking for the "active" flag in the dataset, the inactive vendor could be filtered out during the testing process.

The IT vendor management process can be complex and there may be instances the suspicious payments can be considered legitimate payments. Perhaps a duplicate vendor master data entry was created, and this was realized by the accounts payable just before the payment was processed. In this case, the accounts payable team may have reversed the payment in the system using the duplicate vendor master data item and may have charged it to the original vendor master data entry in the system. This reversed and corrected entry may be hard to identify and comprehend when the IT auditor performed their testing. Hence, the payment may have been flagged as a duplicate payment by the auditor, even though this was not what actually happened. In such situations, the flagged payments can only be confirmed by the accounts payable staff to be legitimate payments.

It could be that the flagged payments were split payments. Some vendors may require that multiple payments be made to other third parties based on their business requirements. For example, supplier ABC may send an invoice with the instruction that 40% of the payment go out to ABC's producers (company XYZ) and 60% made out to supplier ABC. These payments may be hard to identify, and they may seem suspicious to someone who is not aware of the business process.

It is also essential to look at other ways to cross check the payment exceptions. For example, one way to cross-check the preceding process is to use a "three-way match." In a three-way match, the corresponding invoice, purchase order, and receiving report is verified to ensure that the details of a purchase match before a payment is issued. This can be very helpful and can prove as a good initial check to verify the suspicious payments before they are forwarded to the accounts payable check.

Human error is one of the main factors contributing to suspicious payments apart from actual fraudulent payments. Even in highly automated big organizations, there may still be processes that cannot be automated and may need to be manually administered by humans. For example, for the accounts payable process, vendors may send invoices in various file formats such as spreadsheets, MS Word documents, and Adobe PDF documents. Vendors may also send the invoices in their own style. A member of the accounts payable team may need to scan this document and enter in the appropriate details such as invoice date, number, purchase order number, etc. into the correct fields in the organization's accounting system. It may not be too far-fetched that the accounts payable staff missed a zero or period while typing in hundreds of invoices per day manually. If a $100 vendor payment amount was incorrectly entered as $1000, this can be a costly mistake. These types of mistakes are unintentional and may not be treated as a fraudulent payment as per the company policy. However, they still need to be accounted for to ensure they are followed upon by the appropriate parties in the organization. More on using machine learning to detect these types of fraud will be discussed in a later chapter.

Figure 8-1 shows a summary of the verification process for suspected payments.

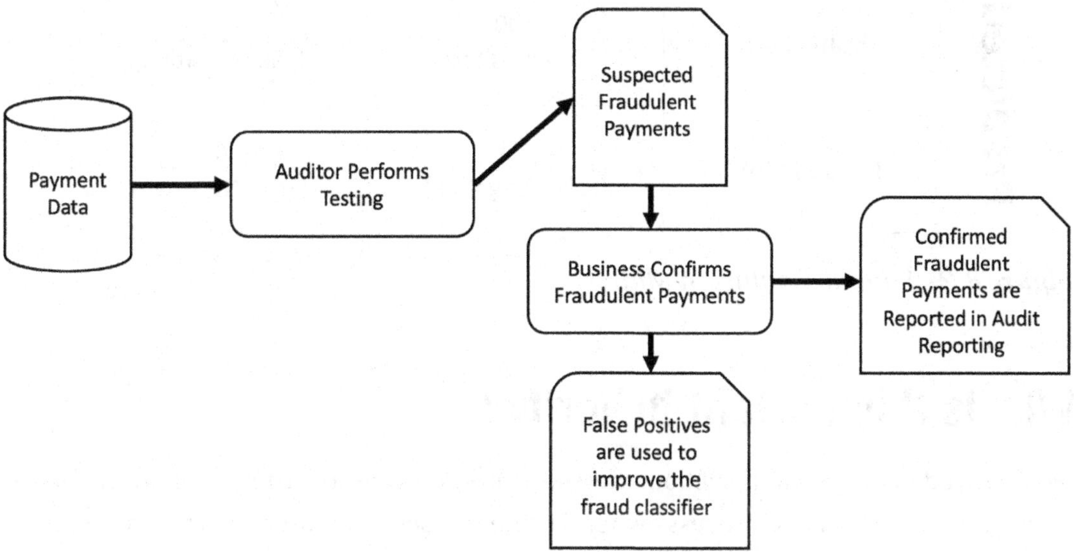


***Figure 8-1.*** *Verification process for suspected vendor payments*

In the preceding example of verifying suspected fraudulent payments, if reversals, payment splits, and human error explain one of the suspicious payments, they can be considered as "False Positives." A false positive is an observation that might look suspicious at first, but it turns out to be legitimate or it can be explained by a business process. A "True Positive," in Figure 8-1, would be an observation that has been confirmed as a fraudulent payment. A "False Negative" would be an observation that was identified as a legitimate payment but was indeed confirmed to be a fraudulent payment. And lastly, a "True Negative" would be an observation that was flagged as a legitimate payment and was confirmed to be a legitimate payment. A confusion matrix, as illustrated in Figure 8-2, captures this information succinctly. A **confusion matrix** is a summary of prediction results on a classification problem. The number of correct and incorrect predictions are summarized with count values and broken down by each class. The most important point to note here is that verification is a highly iterative process.

Figure 8-2 illustrates "False Positives," "False Negatives," "True Positives," and "True Negatives" for 200 observations.

| | | Actual | |
|---|---|---|---|
| | n = 200 | Positive (1) | Negative (0) |
| Predicted | Positive (1) | 100 (True Positive) | 20 (False Positive) |
| | Negative (0) | 10 (False Negative) | 70 (True Negative) |

***Figure 8-2.*** *Sample Confusion Matrix*

## Why Is It Important to Verify?

As discussed in the previous section, there are many reasons to ensure we confirm the findings with the business process owner. Business process requirements and human error need to be taken into account to ensure that false positives are identified and removed from the list of exceptions. Due to its complexity, data science and machine learning applications may pose other unique challenges when it comes to verifying the findings with the business. It may be especially hard to explain how the findings from a machine learning algorithm were calculated. For example, suppose fraudulent payments were identified through clustering. It may be hard to explain why the payments ended up being outliers. The data science professional will need to find a way to explain how the k-means clustering algorithm was used to separate the outlier payments. This is harder than it looks because the level of explanation depends on the experience and background of the person receiving the knowledge. Storytelling is an essential tool when it comes to breaking down complex data into simple digestible information. In the preceding example, storytelling can be leveraged along with evidence from the business to confirm that the suspicious payments were indeed fraudulent payments. This can be a powerful way to explain to various stakeholder groups on the importance of the findings.

For data science and machine learning applications in the area of risk management and auditing, the "Trust, but Verify" principle becomes even more important. Auditing is all about verification and ensuring that the standards set out are being met by the organization's business processes and that the risks are being assessed. Verification serves to support conclusions arrived at over the course of the project. There are many recommended guidelines, according to the Institute of Internal Auditors (IIA), when it comes to audit reporting. However, one of the key aspects of reporting findings is to capture as much information as possible to be able to re-perform the tests and procedures. The idea is that anyone performing the audit process should get the same results shared in the report (provided they have access to the data being tested).

It is important to understand the life cycle of an audit in order to see why verification is crucial in audit applications. The five main phases of an audit conducted by an internal audit are

- **Risk Assessment.** In this collaborative phase, the risks are assessed on an annual basis. The top risks affecting the business process will be discussed and agreed upon by business process owners or leaders. The risks affecting the organization can also be gathered from outside sources such as emerging risks and feedback from other organizations.
- **Audit/Engagement Planning.** The audits will be planned in this step including the formation of the audit team, the people that are qualified to perform the audit, and details on how the audit will be conducted.
- **Fieldwork.** The audit will be conducted in this phase. An audit program is developed based on the risks identified by the business process. Testing procedures will be listed in the audit program in a great level of detail to ensure that they can be re-performed by another auditor.

- **Reporting.** The findings of the audit testing will be shared with the auditees, audit committee, and other management personnel in this step. The corrective action plan will also be listed with the detailed findings. A corrective action plan shows the steps that will be performed in order to overcome a specific problem in order to mitigate the risks.
- **Audit Follow-up.** In this phase, the corrective action plan is checked on a regular basis by the audit contacts to ensure that the recommended items are being closed appropriately.

Figure 8-3 shows the main phases of an audit life cycle.

***Figure 8-3.*** *Phases of an audit*

Almost all of the audit phases in Figure 8-3 are driven through the principle of verification by looking at facts, not just relying on trust alone. Verification is a core theme in audits and is done through checking for evidence to ensure the risks introduced by business processes are appropriately assessed and corrective action plans are in place.

Another complexity introduced by machine learning projects is the randomness of the algorithm. In some of the algorithms such as random forest, independent results may be combined to form a final result, based on randomness. Randomness in machine learning can also occur during data preparation, sampling, cross validation, or weight initialization. What this means is that machine learning depends on randomness to a certain extent to make decisions. Even with the same dataset, multiple results can be obtained from the same machine learning algorithm. This can be a problem when machine learning is applied for audits.

For audit testing, we need the machine learning models to be reproducible so that another auditor can get the same results with the same dataset. A "Random Seed" can help solve this problem. A random seed is the starting point for the random sequence and, if you use the same seed for each ML experiment, the same sequence of numbers is guaranteed. This can easily be done with a few simple lines of code. In Python, you can use the following code to set the random seed to "1000."

```
import random
random.seed(1000)
```

# Integrated Reporting

The standard sections of the reporting audit phase will be shared in the following. This will be followed by the standard deliverables from a data science project. Finally, the sections will be integrated with each other to ensure they support verification of the results.

For an internal audit project, the following sections are included by default for reporting purposes:

- **Executive Summary.** This is the high-level summary of the findings with areas of improvement clearly identified. Some audit shops also highlight areas where the audited business process performed strongly in this section.
- **Key Findings.** In this section, the key findings or exceptions from conducting the audit is listed in one or two sentences. It also captures the risks associated with the findings.

- **Detailed Findings.** In this section, a deep dive into the findings is described in as much detail as possible. The level of details depends on the organization requirement and guidelines. Some organizations capture the supporting data in this section, while others may choose to exclude them due to inclusion of personally identifiable information. The recommendations that need to be taken to close the findings is also listed in this section.

The above is by no means a complete listing of all the sections, but these are the main sections shared in an audit report. In addition to the preceding sections included in audit reporting, audit tests and procedures are captured during the "Fieldwork" phase of the audit. The audit testing step captures all information to the level of detail that anyone who has access to the data would be able to re-perform the tests from.

Here are the major sections of a data science/machine learning report structure:

- **Executive Summary.** This is the high-level summary of the core findings of the project.
- **Introduction.** This section will describe the background and literature review of the methods that will be used in the project.
- **Methodology.** In this section, the data sources and methods used in the project will be explained.
- **Results.** This section presents the findings in terms of descriptive statistics and supported by graphical illustrations (visualizations).
- **Discussion.** In this section, the numbers and illustrations from the previous section will be described in detail. Impactful stories that encourage the reader to take actions will be produced in this section.
- **Conclusion.** The major findings are generalized and re-iterated at a higher level in this section.

Other sections may exist depending on how formal the data science project needs to be captured. Some of the other sections include references and acknowledgement.

The reporting sections from a data science project can be captured as part of the internal audit reporting. Figure 8-4 shows a sample integration of data science reporting sections with audit reporting sections.

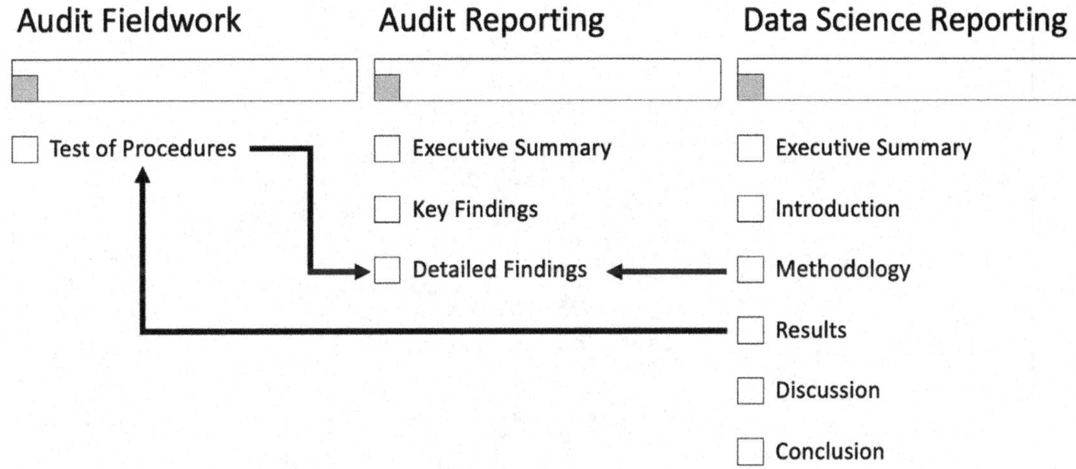


***Figure 8-4.** Sample audit reporting for data science and ML projects*

The methodology sections from the data science project reporting can be captured in the Detailed Findings section of the audit report. The results section showing the details of the ML algorithms and the parameters used can be integrated with the audit testing and procedures section during the Fieldwork audit phase. Supported visualizations can also be included to explain the findings. For instance, it may be easier to convey the findings from a k-means clustering by showing how the clusters were divided by the algorithm based on the features.

Figure 8-5 shows a sample output from a clustering machine learning algorithm. This can be embedded in the final report to show how the clusters were divided and convey where machine learning was used in the audit project.

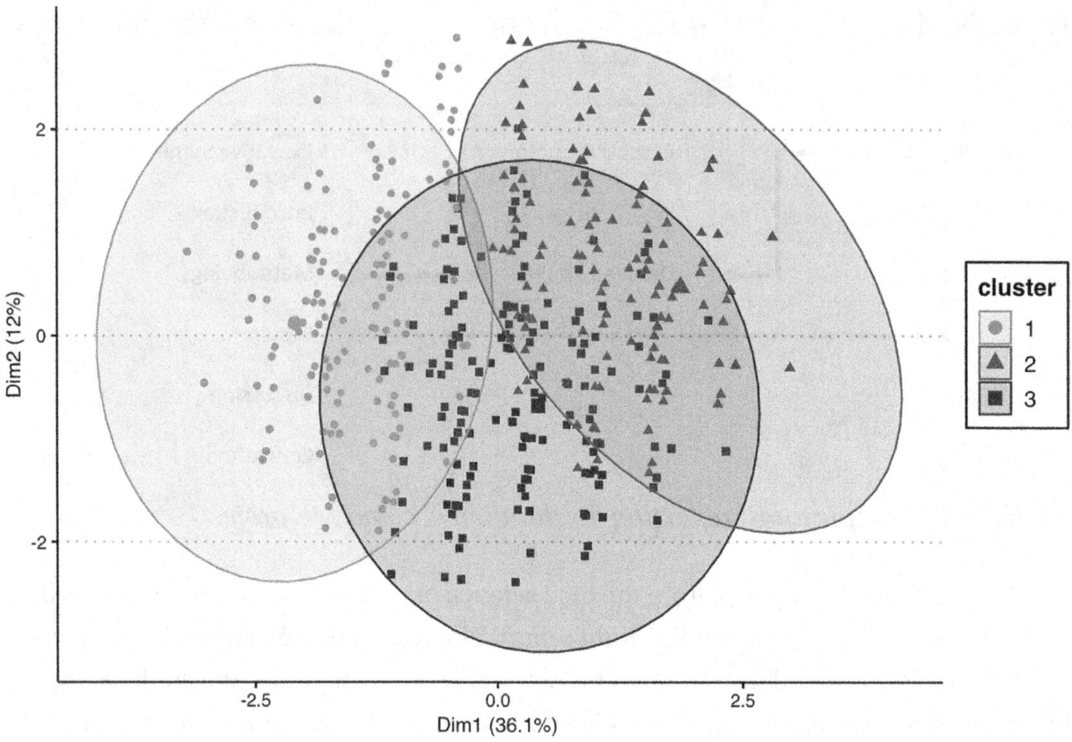


***Figure 8-5.** Sample clustering output*

During the audit reporting phase, the audit report is sent to management (business process owners) with the findings. Management confirms the findings and agrees on the corrective action plans. If the management team does not agree with any of the findings, management will capture this and give their reasoning for the disagreement. For machine learning or data science projects, there may need to be some additional touchpoints with the business to ensure they understand how the data was tested by the audit team. It may be beneficial to both auditors and business process owners to let management know during the planning phase of the audit about the level of testing involved and the level of support needed to confirm the findings.

# Conclusion

Audit findings that involve machine learning and data science projects can be driven using a standard data science/machine learning reporting structure. The Methodology section from the data science reporting structure can be reported in the Detailed Findings section of the Audit Report. The concept of "Trust, but Verify" will need to be exercised when it comes to testing and verifying findings, especially when using data science and machine learning. Findings will always need to be confirmed by the business in order to ensure the findings are legitimate.

# CHAPTER 9

# Machine Learning Fundamentals

In this chapter, the foundations of the machine learning process will be described in detail. Supervised and unsupervised machine learning will be explored further. Some of the most commonly used machine learning models and algorithms will be described in detail. This includes dimensionality reduction. The chapter will also look at some of the things to keep in mind during the feature engineering stage. It also talks about model fitting, specifically, overfitting and underfitting and how to avoid them.

## Supervised Learning

Supervised machine learning algorithms are composed of two distinct types of predictions - classifiers and regression. Supervised machine learning involves the prediction of a specific target variable per observation of results. For example, consider the prediction of whether an observation constitutes a fraud payment. The features used to train the machine learning algorithm could be the payment date, payment amount, vendor details, etc. These features represent the non-dependent variables, and the prediction of whether the payment is a fraud payment or not denotes the dependent variable or target variable to predict. Once the features have been "scrubbed," the features and the training observations can be classified manually by the accounts payable team as a confirmed fraudulent payment. The features and the fraudulent payment indicator can then be used to train the machine learning classifier model.

After the model is trained and validated with another set of data points, it can then be deployed in production to predict fraudulent payments. This example was based on the fact that the output being predicted or target variable takes on only one of two possible categories - "Fraud" or "Not Fraud." The same example can be setup to show the level of confidence we have that a given observation can be a fraudulent payment.



© Maris Sekar 2022
M. Sekar, *Machine Learning for Auditors*, https://doi.org/10.1007/978-1-4842-8051-5_9

This can be done with a machine learning regression model (Neural Networks, for instance). In this case, the predicted output can take on a decimal value from 0 to 1, with a 0.35 indicating a 35% chance that a given observation occurs. In general, if the output being predicted is a categorical variable, a classifier is used. And if the target variable is a quantifiable number, a regression model is used.

---

In general, if the output being predicted is a categorical variable, a classifier is used. And if the target variable is a quantifiable number, a regression model is used.

---

In this book, we will explore the high-level workings of the critical machine learning algorithms needed to solve problems in risk management and auditing. It may not be necessary to go through how the algorithms work in detail. However, a high-level understanding of what the algorithms do and when to use them may be good enough to apply machine learning methods in the most practical way possible. It is the opinion of the author that this is crucial to ensure that machine learning is adopted in an industry setting by end users.

## Classifiers

Classifiers are an essential part of supervised learning as they can be used when predicting categorical variables, as shown earlier. The steps to build a classifier model include the following:

- A classifier is trained with training data (features and the expected output).
- The classifier uses these data observations to identify relationships and patterns of how the observations can be mapped to the expected output.
- After the classifier is trained, a classifier model is generated and evaluated for accuracy through the confusion matrix and ROC curve. A Receiver Operating Characteristic (ROC) curve is a plot that shows the performance of a classifier system.

- The finalized model is then deployed into production. When the model processes new observations, the model will predict the expected output based on the features.

Classifiers are typically used in two different ways - to provide insights into the underlying data and to classify an observation into the correct category. The second application is more commonly used. The preceding example on the fraud payments classifier is an excellent example of the second application. Another way the preceding example can be used with a classifier is to understand what patterns contribute to fraud. In the preceding example, after classification is performed, the confirmed fraudulent group can be further analyzed to find the combinations of features that create a more conducive environment for fraud. Do most fraud payments occur on a specific day (perhaps on a Friday?) and with a numbered company (149324 PROVINCE LTD.)? This type of question can give additional insights into the data.

## Decision Trees

A decision tree is a type of classifier that is simple to understand. It follows a tree structure with nodes and leaves (at the lowest level). Each node consists of a question about one feature. If the feature (at the node level) is categorical, there will be different child nodes corresponding to each of the categories. On the other hand, if the feature is continuous, a threshold is used to get a "Yes" or "No" response for the corresponding child nodes. At the lowest level, each leaf shows the proportion of the splits in the target variable being predicted.

Although it is not necessary to understand the exact steps taken to create a decision tree, a brief explanation will be given for reference. Here are the steps:

- Find the feature that best partitions your data into the expected classes. For instance, in the preceding example, the classes would be "Not Survived" and "Survived." This will become the root of the tree.
- Recursively, train each child node by partitioning the best feature that partitions the data into the classes.
- Once all features are exhausted or the maximum depth (parameter) is reached, the algorithm stops leaving the leaf nodes with the partitioned classes along with their proportions.

Information gain or Gini impurity are most likely used to determine the proportions at each tree level. These are specifics that can be read upon further, but will not be discussed here.

The Python library Sklearn can be used to create a decision tree classifier. Here is sample code to create a decision tree classifier and train it based on the sample iris dataset in Python:

```
from sklearn.datasets import load_iris
from sklearn import tree
iris = load_iris()
x, y = iris.data, iris.target
clf = tree.DecisionTreeClassifier()
clf = clf.fit(x, y)
```

In the preceding code, x represents the training data and y represents the expected output. After training the decision tree classifier, we can plot to see the tree using this code:

```
tree.plot_tree(clf)
```

We can predict using the trained decision tree classifier with the following code:

```
print(clf.predict([[1., 2., 3., 4.]]))
```

Figure 9-1 shows the decision tree created for the sample iris dataset provided within Python. Here, X[0] is the first feature, X[1] is the second feature, and so on.

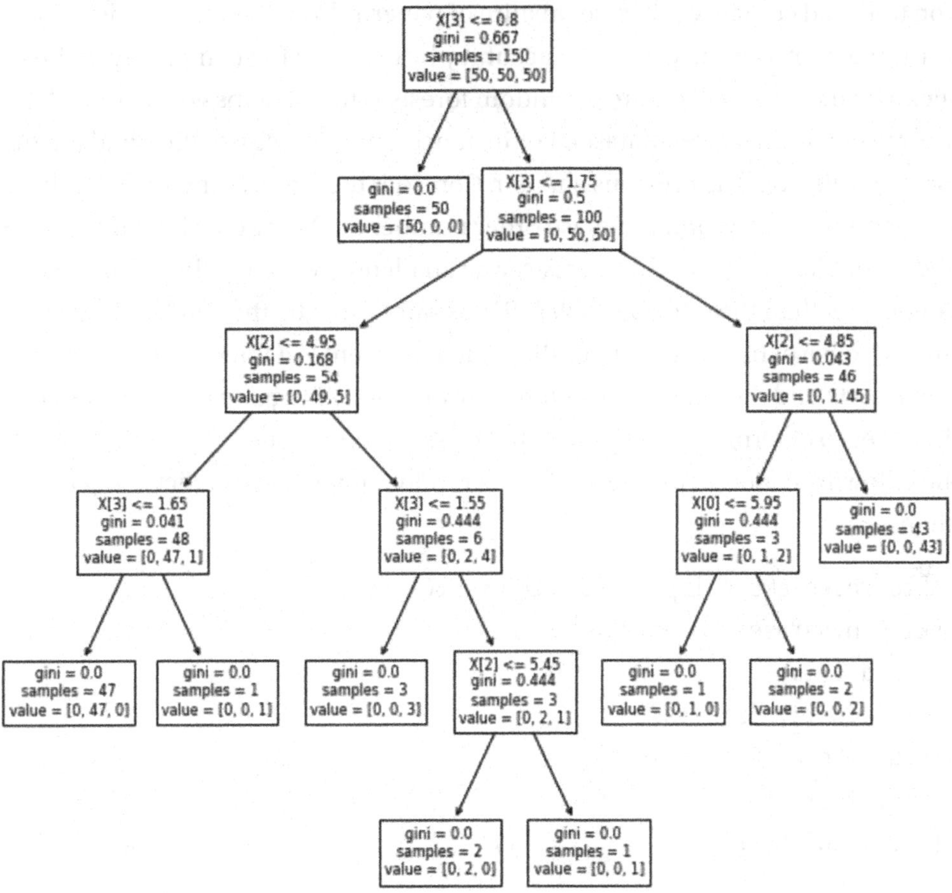


***Figure 9-1.** Decision tree of the iris dataset*

Decision trees divide the data set into two sections at every node based on the condition set in the node. For example, the root node in Figure 9-1 checks if the fourth feature is ≤ 0.8. The data is divided based on this condition.

# Random Forests

Random forests utilize decision trees in their algorithm. A random forest is an ensemble learning method that operates by constructing a multitude of decision trees at training time for classification, regression, and other machine learning tasks. For classification tasks, the classification score of each of the individual decision trees is averaged to get the classification score of the trained random forest model. Random forests typically perform better than decision trees as classifiers are not "pruned," unlike a single decision tree, and contains a diverse mixture of patterns from the random decision trees. Pruning improves the predictive accuracy by reducing the complexity of the final classifier. The most crucial feature of random forests is that it helps get an idea of the feature importance in a given dataset. For instance, suppose we use the fraud payment analysis example from the previous chapter. Consider the features invoice date, invoice number, vendor country, and purchase order number are being considered along with ten other variables. Using random forests, we can identify which of these features are more important than the other variables. For example, maybe the vendor country is determined to have the most say in finding out a payment is a fraudulent payment. Random forests find the feature importance by randomly plugging a feature into the decision trees to determine its effects on the classification score.

The following code can be used to train a random forest and predict using it in Python:

```
from sklearn.ensemble import RandomForestClassifier
from sklearn.datasets import load_iris
iris = load_iris()
x, y = iris.data, iris.target
clf = RandomForestClassifier()
clf = clf.fit(x, y)
print(clf.predict([[1., 2., 3., 4.]]))
```

## Support Vector Machines

Support Vector Machines (SVM) is a classifier algorithm that relies on the concept of linear separability. With SVM, there are two classes separated by a plane (3-dimensional) or line (2-dimensional) such that all data points belonging to the first class are separated from the data points belonging to the second class. An SVM is effective if data can be logically split into two classes. However, this may not be the case for all datasets in reality. It is an efficient algorithm if it does appear that the dataset can be clearly split with a plane (or line). In higher dimensional space (when there are many features), dimensionality reduction algorithms can be leveraged to reduce the dimensionality of the dataset for SVM use.

Figure 9-2 shows the concept of linear separability used by the SVM algorithm.

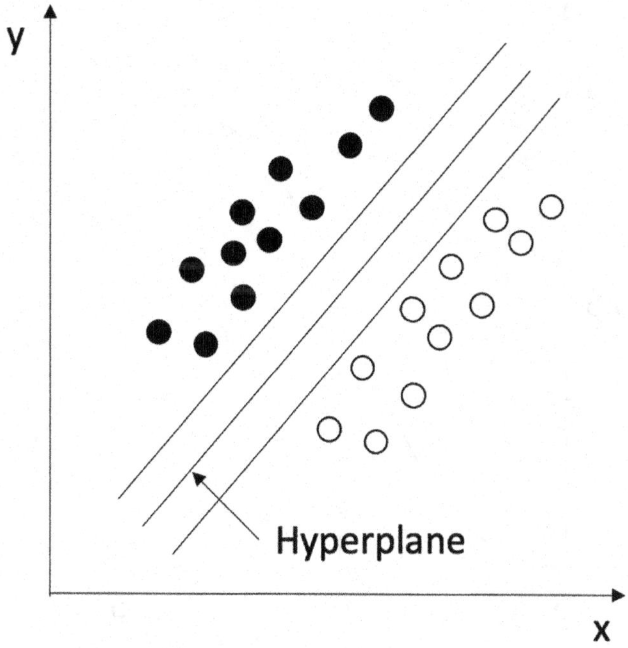


***Figure 9-2.*** *Linear separability*

The following code can be used to train an SVM classifier:

```
from sklearn.linear_model import SGDClassifier
from sklearn.datasets import make_blobs
X, Y = make_blobs(n_samples=50, centers=2, random_state=0)
clf = SGDClassifier(alpha=0.01, max_iter=100)
clf = clf.fit(x, y)
```

## Logistic Regression

A logistic regression classifier model works similar to an SVM model, but it gives a non-binary output. Scores are assigned based on how far the points are from the hyperplane. A hyperplane divides an n-dimensional space into two disconnected parts. A sigmoid function is used to derive the score. This function is set up in such a way that the score is 0.5 at the hyperplane and goes towards 0 or 1 if the distance from the hyperplane is decreased or increased. The function forces the score to have a minimum of 0 and a maximum of 1.

Figure 9-3 shows the sigmoid function.

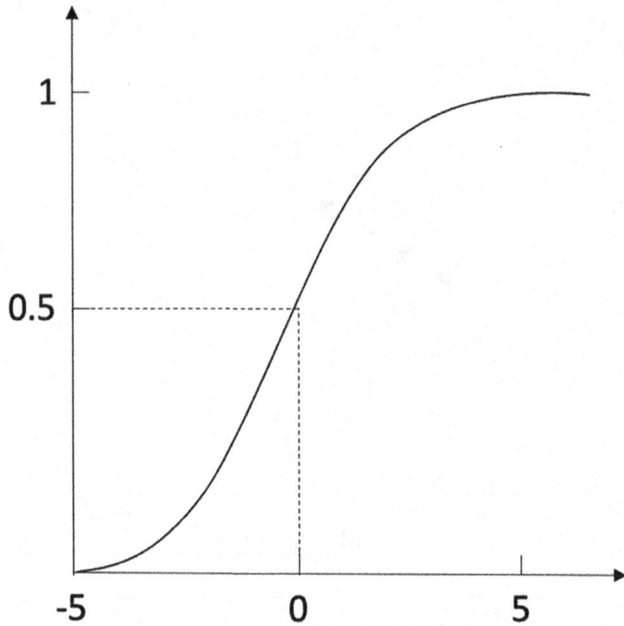


***Figure 9-3.*** *Sigmoid function*

The following code can be used to train a logistic regression classifier:

```
from sklearn import linear_model
clf = linear_model.LogisticRegression()
clf.fit(x, y)
```

One of the problems of logistic regression is that most features may have weights (coefficients) that are small with no clear differentiation between them. For the classifier to work optimally, we need fewer features with large weights that can provide some insights into the data. A lasso regression helps with this by providing a way to map most feature weights to zero and only the features that are important have non-zero weights.

## Naive Bayes

Naive Bayes is a class of supervised machine learning problems that are based on the Bayes' theorem with the assumption that the features have conditional independence. If two events are conditionally independent, the first event does not affect the outcome of the second event. The specifics of Naive Bayes' classifiers will not be discussed here. What is interesting to note for this classifier is that initially a confidence is assigned to each of the classes (e.g., 0 or 1 in the case of a binary classification problem). The confidence is then changed as more information becomes available. It can be seen how a wide range of real-world applications are possible as a result of this algorithm. Even though there is an assumption of conditional independence between the features, the algorithm is surprisingly powerful in real-world applications with a high degree of accuracy.

The following code can be used to train a Gaussian Naive Bayes classifier. In this classifier, the likelihood of the features is assumed to be a Gaussian distribution (the bell curve):

```
x_train, x_test, y_train, y_test = train_test_split(x, y, test_size=0.1)
gnb = GaussianNB()
y_pred = gnb.fit(x_train, y_train).predict(x_test)
```

Figure 9-4 shows a normal distribution (an example of a Gaussian distribution).

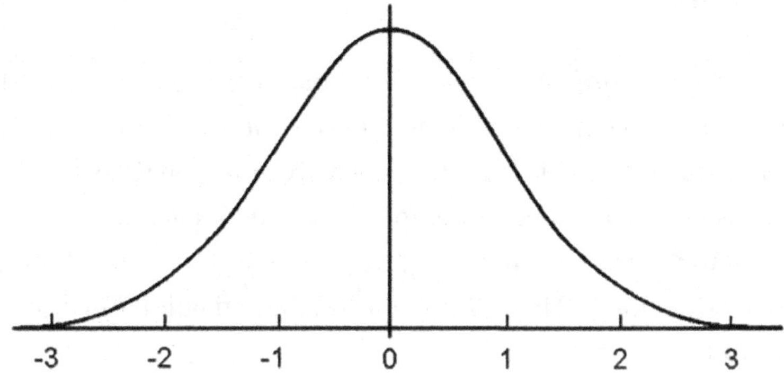


***Figure 9-4.*** *Normal distribution*

## Deep Learning

Artificial Neural Networks (ANN), or simply neural networks, have been widely used in machine learning for classification and regression problems. More recently, deep learning, a subfield of neural networks, has become more popular for its ability to scale and perform better with large datasets. Neural networks have been inspired by biological neural networks that are seen in the human brain. A neural network consists of a collection of connected nodes called neurons, similar to how neurons work in the human brain. There are many different types of neural networks. A perceptron is the simplest form of neural network. Each neuron in a perceptron takes multiple inputs and gives out a single output.

Figure 9-5 shows the perceptron consisting of the interconnected nodes.

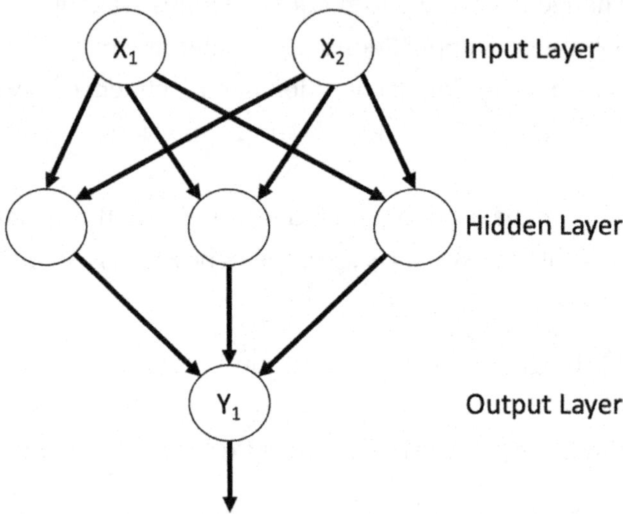


***Figure 9-5.*** *Perceptron*

In Figure 9-5, each neuron is activated using a function called the activation function. An *activation function* of a node defines the output of that node given an input or set of inputs. There are different types of activation functions for various applications. If the sigmoid activation function is used, then each neuron can act as a logistic regression function as seen in the logistic regression section earlier. Each neuron takes a weighted combination of inputs and applies the activation function in order to determine the output.

As shown in Figure 9-5, the starting layer takes in all features as inputs. So, the starting layer represents the input layer. The last layer at the bottom is the output layer. All the layers in between are known as hidden layers. When there are many hidden layers, the neural networks are considered deep networks. Training deep networks is where the term deep learning comes into play.

The following code can be used to train and predict using a multi-layer perceptron classifier:

```
from sklearn.neural_network import MLPClassifier
clf = MLPClassifier(max_iter=100).fit(x_train, y_train)
clf.predict(x_test)
```

# Confusion Matrix

Supervised learning algorithms that classify data were discussed in the previous sections. In this section, the evaluation of classification models will be explored.

The next logical step in building a machine learning model is to find out how accurately the model performs with the dataset. False positives and false negatives need to be identified to get an idea of the performance of the classifier. When the predicted outcome is different from the actual result, false positives and false negatives are encountered. For a binary classifier, false positives occur when the classifier predicts the outcome to be a 1, but the actual result was a 0. Likewise, false positives occur when the classifier predicts the outcome to be a 0 when the actual result was a 1. Based on the business problem at hand, the false case (positives or negatives) that is of higher importance is determined.

When a classifier is designed to detect cancer patients, the classifier needs to identify as many cancer patients as possible – thus, the classifier needs to flag 1s more aggressively so a false positive may be more acceptable. If a classifier is built to invest in projects based on project variables, the classifier will need to flag 1s more conservatively to stay within the budget.

Metrics such as precision, recall, true positive rate, and false positive rate (among others) can be used to measure the performance of a classifier. Here are some high-level key points for these metrics:

- **Precision.** This shows the proportion of positive identifications that ended up being correct.
- **Recall.** This shows the proportion of actual positives that were flagged correctly.
- **True positive rate.** This is the same as precision. If true positive rate is 1.0 then 100% of the flagged positives were correct.
- **False positive rate.** This is the proportion of positives that should not have been flagged. A false positive rate of 0 implies that all positives were correctly flagged.

A confusion matrix can be used to represent the number of data points that fall in each of the categories: true positives, true negatives, false positives, and false negatives.

Figure 9-6 shows the confusion matrix for a binary classifier where either a 0 or 1 are the only outcomes.

| n = 200 | Actual: Positive (1) | Actual: Negative (0) |
|---|---|---|
| Predicted: Positive (1) | 100 (True Positive) | 20 (False Positive) |
| Predicted: Negative (0) | 10 (False Negative) | 70 (True Negative) |

***Figure 9-6.*** *Confusion matrix for a binary classifier*

The following Python code can be used to build a confusion matrix:

```
from sklearn.metrics import confusion_matrix
y_true = ["0", "1", "0", "0", "1", "1"]
y_pred = ["1", "1", "0", "1", "0", "0"]
tn, fp, fn, tp = confusion_matrix(y_true, y_pred, labels=["0",
"1"]).ravel()
(tn, fp, fn, tp)
```

In the preceding code, tn, fp, fn, and tp represent true negatives, false positives, false negatives, and true positives, respectively.

## ROC Curves

A **Receiver Operating Characteristic (ROC)** is a graph that shows the performance of a classifier at all thresholds for the true positive rate and the false positive rate.

The true positive rate vs. false positive rate is plotted, as shown in Figure 9-7. Here, a classifier that has a lower false positive rate and higher true positive rate is preferred, since this means the classifier model is able to predict far more items correctly. Both axes range from 0 to 1. An ROC can be used to see the performance of multiple classifiers or multiple classes of a single classifier.

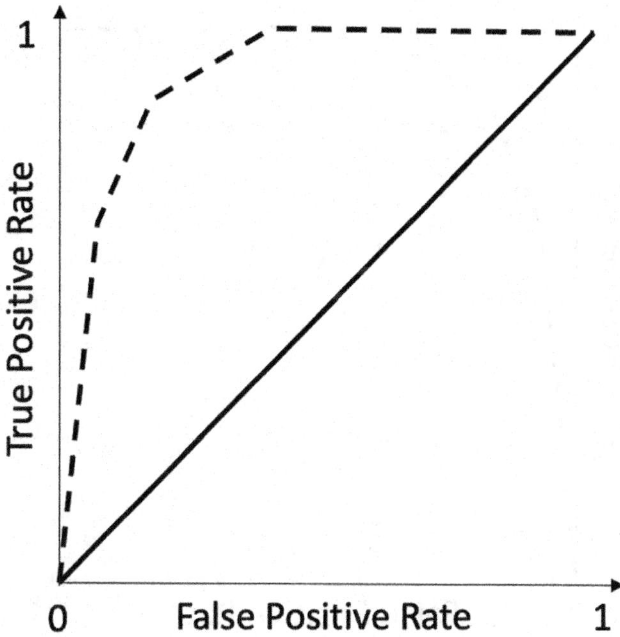


***Figure 9-7.*** *ROC curve*

In Figure 9-7, the area under the ROC curve (AUC) can be calculated to get the aggregate measure of performance across all thresholds. Figure 9-8 shows the AUC as the shaded portion under the curve. AUC ranges from 0 to 1 with 0 meaning that 100% of the predictions were wrong and 1 meaning that 100% of the predictions were correct.

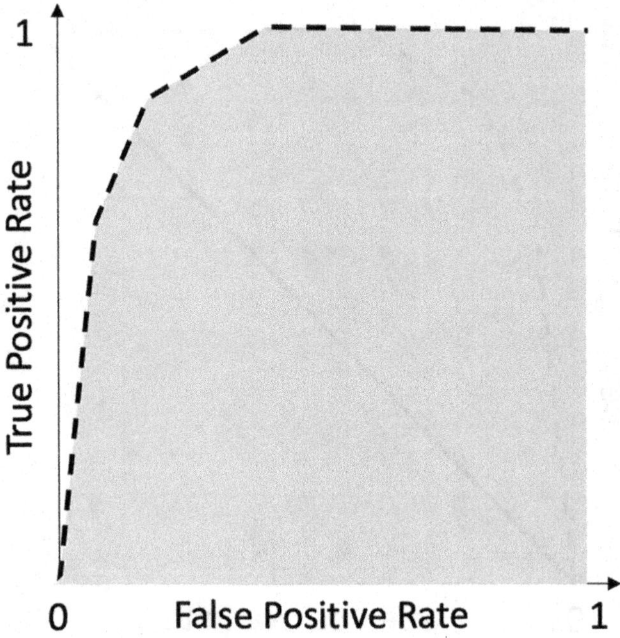


***Figure 9-8.*** *Area under ROC curve*

The following code can be used to plot the ROC and calculate the AUC:

```
from sklearn.metrics import roc_curve, auc
from sklearn.metrics import roc_auc_score
fpr, tpr, _ = roc_curve(y_test, y_score)
plt.figure()
lw = 2
plt.plot(fpr, tpr, lw=lw,
      label='ROC curve (area = %0.2f)' % roc_auc)
plt.plot([0, 1], [0, 1], lw=lw, linestyle='--')
plt.xlim([0.0, 1.0])
plt.ylim([0.0, 1.05])
plt.xlabel('False Positive Rate')
plt.ylabel('True Positive Rate')
plt.title('Receiver operating characteristic example')
plt.legend(loc="lower right")
plt.show()
roc_auc = auc(fpr, tpr)
```

## Regression

**Regression** is a set of supervised machine learning techniques used to predict a continuous output variable based on a set of input features. At a high level, the coefficients of the input features are determined so that the output variable can be shown in terms of the input features. Once the coefficients are solved for, they can be plugged into an equation and used to predict the output for other sets of input variables.

## Linear Regression

In linear regression, it is assumed that the features have a linear relationship with the output. The linear regression equation can be defined as

$$y = b_0 + m_1x_1 + m_2x_2 + m_3x_3 + \ldots + m_ix_i + e$$

where y is the output variable,

$b_0$ is the intercept,

$m_i$ are the coefficients,

$x_i$ are the input features used for prediction,

and e is the residual error.

A method known as the Ordinary Least Squares (OLS) can be used to calculate the coefficients. OLS will not be covered in this book.

The most popular way to evaluate regression models is to use Root Mean Squared Error (RMSE). The Mean Squared Error (MSE) is the mean of all the differences between the predicted output ($y_i$) and actual output squared. The RMSE is obtained by taking the square root of the MSE.

The lower the RMSE, the better the regression model is performing. A higher RMSE means there is a bigger difference between the predicted and the actual outputs, which means the regression model is performing poorly.

The following Python code can be used to build a linear regression model and calculate the RMSE:

```
from sklearn.linear_model import LinearRegression
reg = LinearRegression().fit(x, y)
reg.coef_
reg.intercept_
reg.predict(xtest)
print(np.sqrt(metrics.mean_squared_error(y_obs, y_pred)))
```

The difference between the logistic regression and linear regression models is that the logistic regression model is a classifier used to classify a binary or categorical output (for example, if a payment is fraudulent or not). The linear regression model can be used to calculate a continuous output (for instance, to predict the next payment to a vendor). Figure 9-9 shows how the output variable varies for logistic regression and linear regression models.

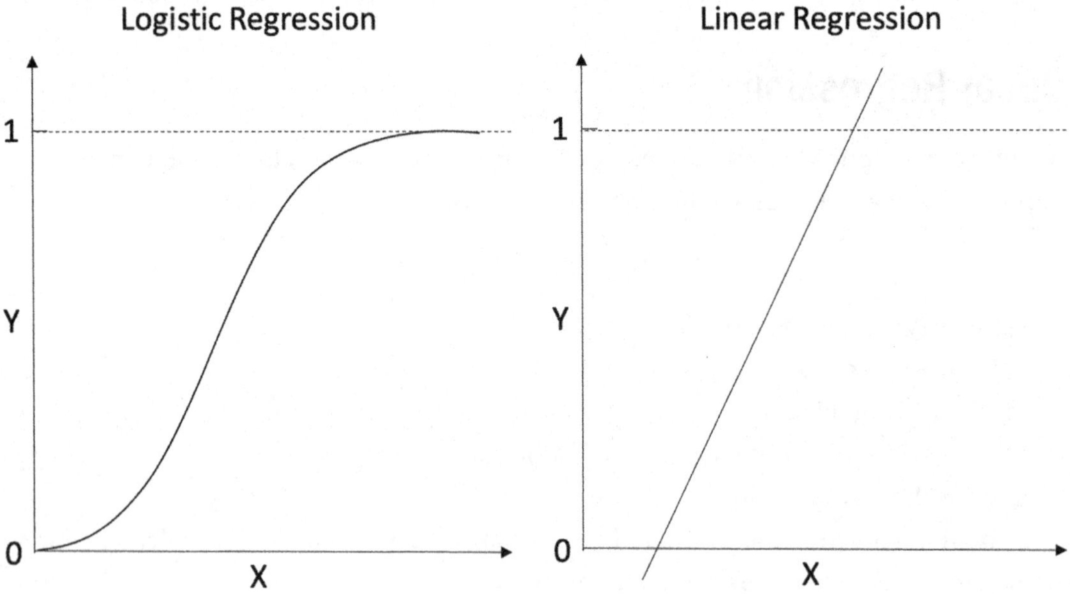


***Figure 9-9.*** *Logistic regression vs. linear regression*

## Unsupervised Learning

**Unsupervised learning** is a group of machine learning algorithms that are used to derive hidden insights from the data without specifying the labels of the features. Clustering is one of the popular applications of unsupervised machine learning. The same example used to identify fraudulent payments made to a vendor can be used to demonstrate unsupervised learning. Instead of identifying the features (payment date, payment amount, vendor details, etc.) and the targeted output (fraudulent payment or not), we can feed in the features into a clustering algorithm. The dataset does not need to be divided into the training set and test set as seen for supervised machine learning.

The clustering algorithm will use the Euclidean distance between the features to come up with clusters based on the distance of the data points from one another.

The number of clusters can be any number based on the business context and the requirements of the projects. There is a way to calculate the optimal cluster size, which will be discussed later in detail. The clusters can then be analyzed with the corresponding features to get insights into outliers seen in the dataset or payment behavior of the company. For example, Cluster 1 may consist of payments made to vendors who were recently added in the system with a higher payment amount. Cluster 2 may consist of vendor payments made to old vendors with a higher payment amount. If there was a payment fraud in the past, the conditions surrounding the fraudulent payment can be analyzed and cross-checked with the clusters to see if any of the clusters exhibit similar feature characteristics. If there is a match, the cluster of payments can be further analyzed to confirm if fraud has taken place. If there has not been an instance of fraud, the characteristics known to increase the chance of fraud such as if the company is a numbered company, or if the company has been in business only for a short time can be used as a way to see if any of the clusters resonate with these characteristics.

# Clustering Algorithms

k-means and hierarchical or agglomerative clustering are two clustering algorithms that are widely used. In this section, we will discuss these two algorithms in detail.

# k-means Clustering

k-means clustering finds groups of data within a dataset that have similar features and assigns them to the nearest cluster centroid. A centroid is an arithmetic mean of all points in the data set. The number of centroids is specified in order for the algorithm to segment the data. Customer behavior segmentation, anomaly detection, and inventory categorization are some common use cases of k-means clustering.

The following code can be used in Python to create a k-means cluster with 3 clusters and plot it with color:

```
from sklearn.cluster import KMeans
kmeans = KMeans(n_clusters=3, random_state=0).fit(x)
```

Once the clusters are identified by the k-means algorithm, they can be joined with the original dataset and plotted to see how the clusters are distributed across the variables.

Figure 9-10 shows a sample k-means cluster showing how the clusters are distributed between two variables. Here, ASD stands for Autism Spectrum Disorder.

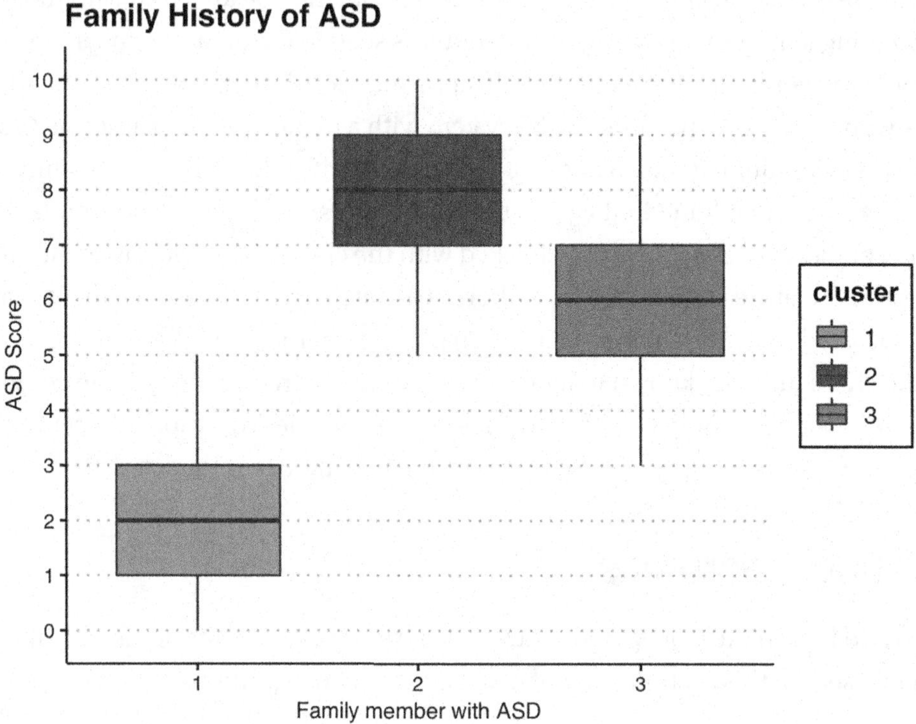


***Figure 9-10.*** *Distribution of clusters*

# Hierarchical Clustering

In agglomerative or hierarchical clustering, a bottom-up approach is used to cluster the data points. For instance, assume the points to cluster are as shown in Figure 9-11.

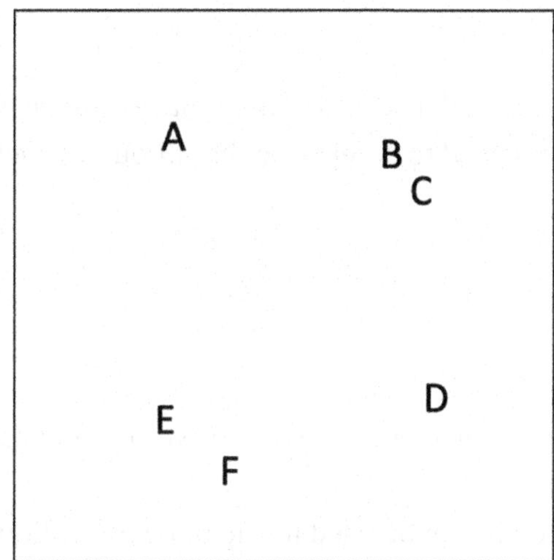


***Figure 9-11.*** *Sample data points*

The resulting clustering is displayed in a hierarchical tree-like structure, which is known as a Dendrogram. A dendrogram for the preceding data points is shown in Figure 9-12.

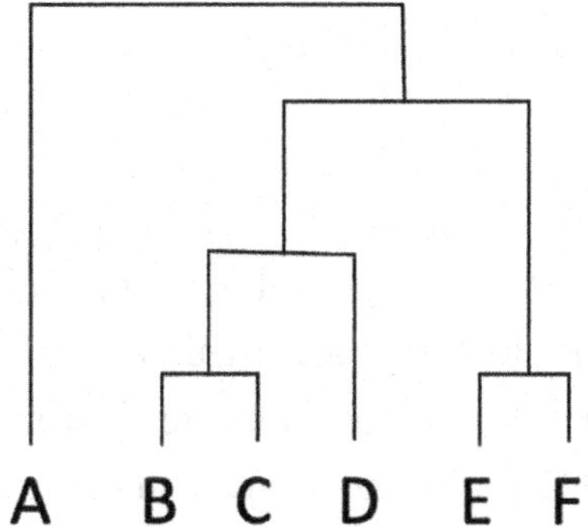


***Figure 9-12.*** *Dendrogram*

The following Python code can be used to perform hierarchical clustering:

```
from sklearn.cluster import AgglomerativeClustering
model = AgglomerativeClustering(distance_threshold=0, n_clusters=None)
model = model.fit(x)
```

In order to evaluate the clustering algorithms, a Silhouette Scores method can be used as a way to determine how geometrically distinct the clusters are from each other.

## Silhouette Score

*Silohouette Scores* measure how dense and widely spaced out each cluster is from one another. A simple formula is used to determine the silhouette coefficient between two clusters:

$$s = \frac{(b-a)}{(a,b)}$$

where a is the average distance between points within a cluster and b is the average distance between all clusters.

The following Python code can be used to calculate the Silhouette Score:

```
from sklearn.metrics import silhouette_score
print(silhouette_score(x, label))
```

The silhouette coefficient is between -1 and 1. If it is closer to 1, the point is closer to points in its own cluster. A score close to 0 means the point is about the same distance from the two clusters. A score of less than 0 means that the point is not in the correct cluster.

## Elbow Plot

An *Elbow Plot* is used to determine the optimal number of clusters for the k-means cluster algorithm. The k-means clustering algorithm is run for k = 1 to 10 and for each k, the sum of squared distances for each point in the cluster from its assigned center is calculated (called distortions). They are all then plotted, and the resulting plot looks like an elbow. The size of the cluster that is closed to where the line bends sharply (at or near the elbow) is considered to be the optimal number of clusters. This is because after reaching the elbow for every increment in the cluster, the y-axis value (distortions) has minimal reduction in its magnitude.

Figure 9-13 shows a sample elbow plot. The optimal number of clusters in this plot is 3 as can be seen in the figure.

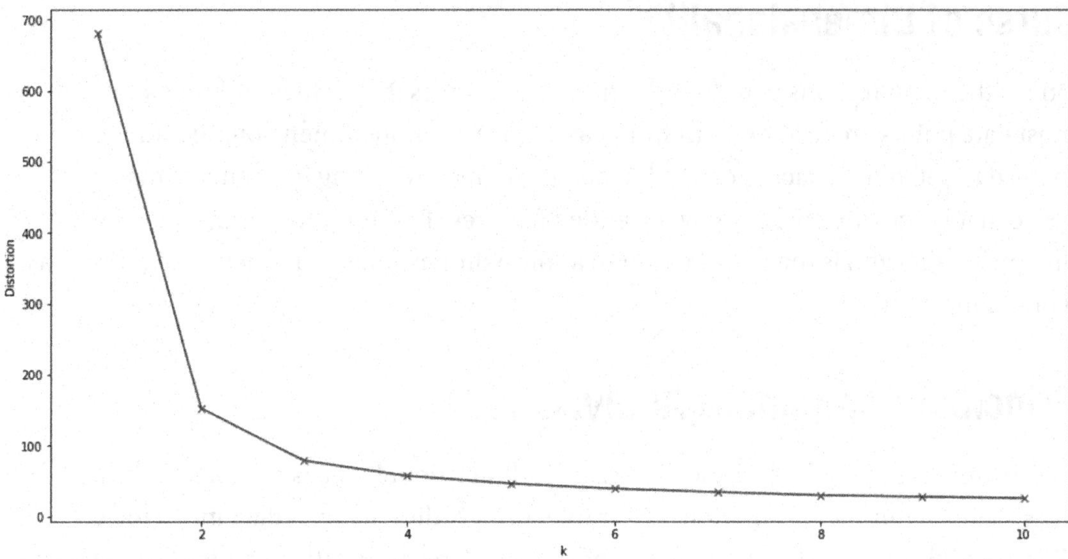


***Figure 9-13.*** *Elbow plot*

# Dimensionality Reduction

Dimensionality reduction is a class of unsupervised learning techniques that deal with reducing the dimensional size of a dataset. The assumption of dimensionality reduction is that for a data set containing d dimensions (features), two or more features can be reduced to one feature in k-dimensional space (where k is much smaller than d). This can be especially useful for processing sound, image, or video files where the number of features can be quite large. For instance, one of the common ways to process images is to read their pixel data to extract meaningful features. Each pixel may consist of 0 or 1 for encoding black and white colors, or more complex, which can compose of one number for red, green, and blue. Suppose human lung x-ray images need to be input into a machine learning algorithm to detect cancer. In these applications, it would be beneficial to be able to reduce the number of features so that the same data can be represented with a smaller subset to improve performance of the algorithm toward detected cancer in patients.

Principal Component Analysis can be used to perform dimensionality reduction and will be discussed in detail in the following text.

## Curse of Dimensionality

One of the main reasons to reduce the dimensionality is that in higher dimensions, most data points are further apart from each other. As more dimensions are added to the dataset, more space is created in between them, resulting in a much smaller space that is actually occupied by the added features. This is known as the curse of dimensionality and is the main reason to apply a dimensionality reduction algorithm as a preliminary step.

## Principal Component Analysis

In Principal Component Analysis (PCA), high dimensional data is assumed to be a big sphere capturing information. PCA projects high-dimensional data into a lower dimensional space so that the data can be captured in a best-fitting line that minimizes the average squared distance of the points to this line. As other similar concepts in this book, we will not delve deeper into this.

Figure 9-14 shows an example output from performing a PCA. In this case, the higher dimensional data is represented in two dimensions, X and Y.

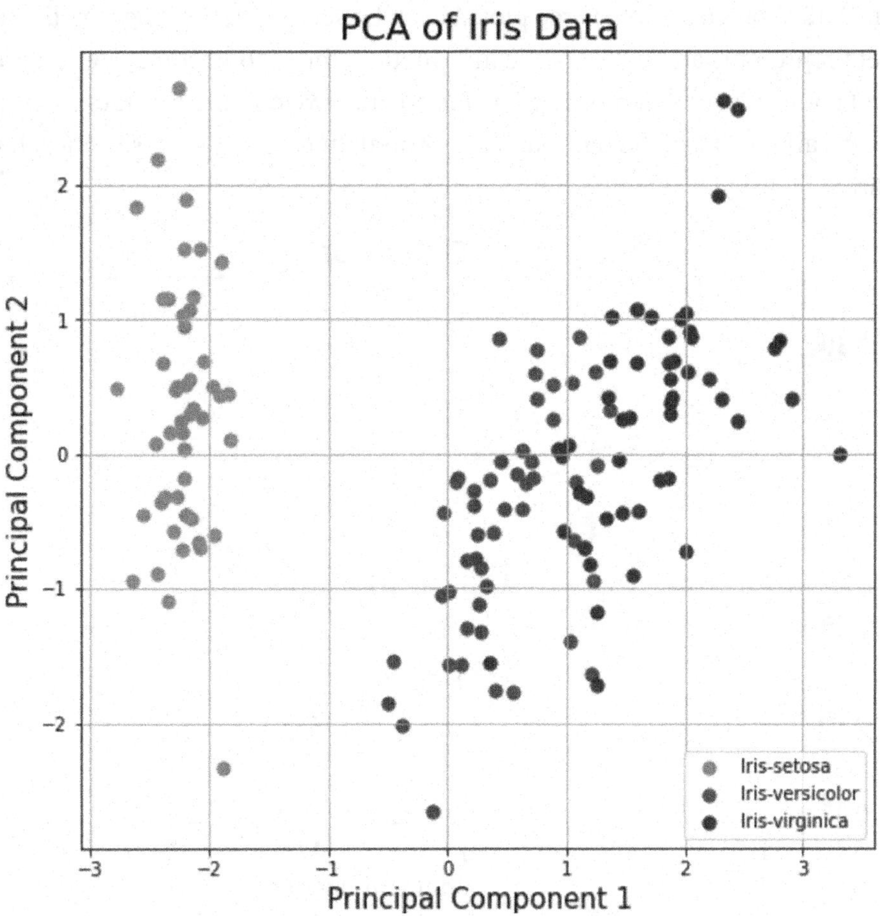


***Figure 9-14.*** *Principal Component Analysis of sample data*

The following code can be used to perform PCA in Python:

```
from sklearn.decomposition import PCA
pca = PCA(n_components=2)
principalComponents = pca.fit_transform(x)
```

## Scree Plots

A Scree plot can be used to see how many principal components are needed to capture most of the information from the dataset prior to PCA being applied.

Figure 9-15 shows a sample Scree plot. As can be seen from the diagram, the first component represents about 30% of the information, the second component represents captures of about 15% of the information, and so on. If we add the first four principal components, about 70% of the data can be captured. In essence, we lose 30% of the information.

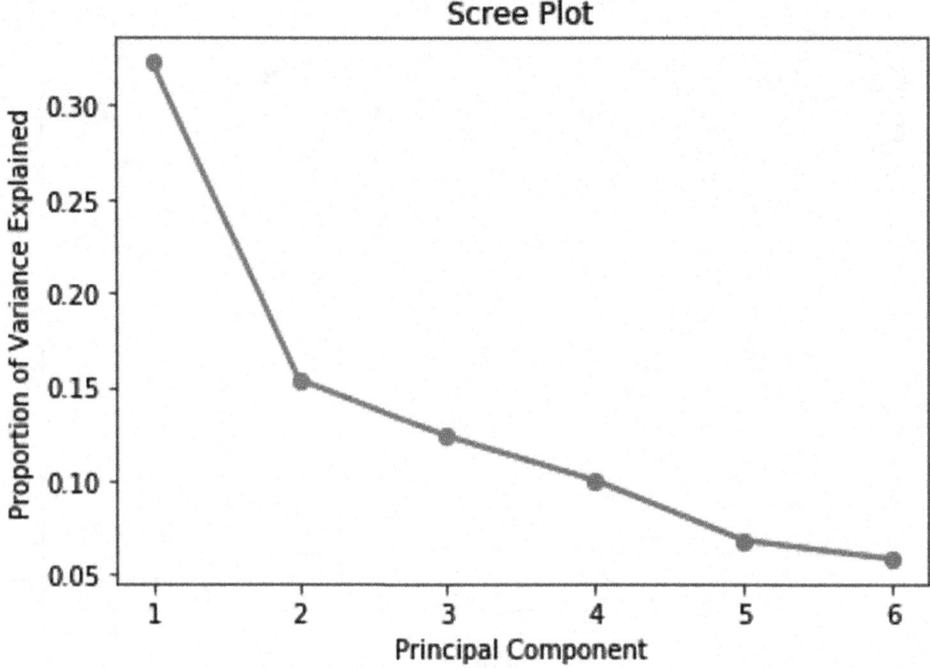


***Figure 9-15.*** *Scree plot*

# Overfitting, Underfitting, and Feature Extraction

The topics in this section are important general guidelines to keep in mind for machine learning model fitting. Overfitting and underfitting are two supervised learning problems that are commonly seen and must be accounted for. Feature extraction is a specialized topic that deals with the transformation of features before they are used by machine learning applications. In many cases, features need to be transformed in such a way that they can be appropriately utilized by the machine learning algorithm.

## Overfitting

Overfitting has always been a major barrier to the effectiveness of machine learning applications. In general terms, overfitting occurs in supervised machine learning applications when the model appears to be effective with training data but does not perform well with test data. For example, suppose we train a supervised machine learning model (classifier) to detect fraudulent payments and include the vendor ID as one of the features. The classifier model may perform well for vendor IDs in the training data, but may not recognize the new vendor IDs in the test data. As a result, it performs well with training data but is not able to generalize the learning and, as a result, the classifier performs poorly with the test data. There are many such instances where overfitting may be an issue.

Here are some of the ways to overcome overfitting:

- Randomize the dataset before it is split between training, testing, and validation. This may be surprisingly effective in reducing bias introduced as a result of taking the first half as training set and second half of the transactions as the test set. In the preceding example, they have had some recurring payments in the training data, but not in the testing data.
- Many supervised machine learning models do not use a validation set. This is especially important when multiple models are being contested and a final model is picked. When multiple models are used, each model is trained using the training data and in order to evaluate the model the testing data is utilized. The final selected model needs to further be tested with a holdout dataset that has never been seen before in order to see the selected model's effectiveness with real-life data.
- k-fold cross-validation is a resampling technique that is used to deal with overfitting in small datasets. It randomly shuffles the dataset into k partitions. Each partition is then used as a test set and the remaining partitions are taken as the training set. The evaluation score is then noted for each of the k partitions and averaged at the end to arrive at the final performance score of the model.

Figure 9-16 illustrates the k-fold cross-validation procedure.

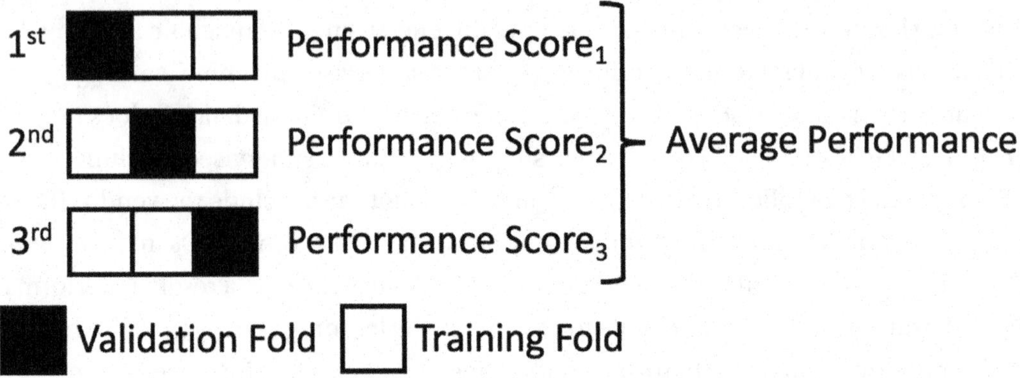


***Figure 9-16.** k-fold cross-validation illustration*

## Underfitting

Underfitting occurs when the supervised learning model is not able to learn from the training set and, as a result, the model is not able to generalize. In the classifier example, this means we have a lot of false positives and false negatives.

The optimal model fitting should have a balanced model that learns just enough to generalize the dataset being used.

Figure 9-17 shows graphs that represent overfitting, underfitting, and balanced model fits. The line represents the model fit with the provided data (circles).

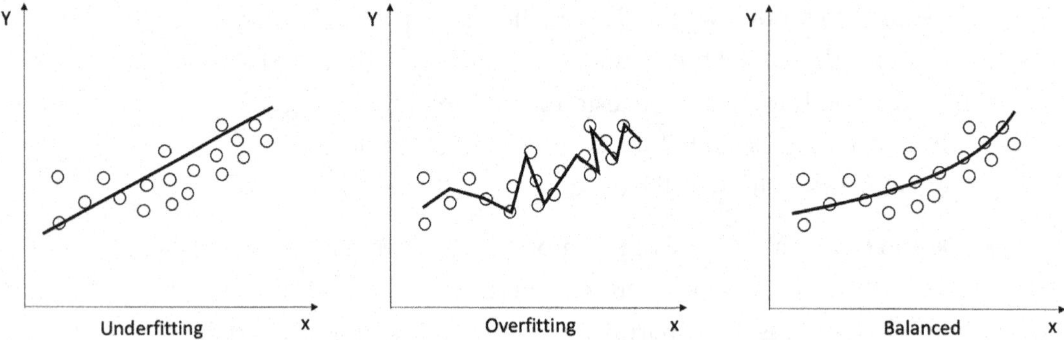


***Figure 9-17.** Model fit illustrations*

## Feature Extraction

Features from a dataset often need to be transformed to be used in a model. That is, they cannot be used in their raw state. Feature viability is as important as the model being selected for use in the machine learning application. In this section, we will look at some common feature extraction situations to keep in mind and how to use them. These transformations are almost always performed during the Exploratory Data Analysis stage (EDA). It can also be performed in other stages depending on findings. For example, during your model evaluation stage you determine that you need to try a different feature that you think might improve the performance.

You might want to create dummy variables that capture only certain categories in a categorical field. Some fields such as payment types may have a large number of categories. Only certain payment categories may need to be really captured as part of the analysis. Another reason to create a dummy variable is to convert a categorical variable to a numeric variable since some algorithms cannot work with categorical data. One method to perform this is through one-hot encoding. In one-hot encoding, a new binary variable is created to capture a "0" or "1" if the category appears in the categorical field. For example, a dummy variable may be created just to flag if the payment type was a vendor payment type. For all observations where the categorical variable indicates a vendor payment, the newly created binary variable would indicate a "1."

In some cases, the variables may need to be transformed from a continuous variable to categories. For instance, age groups of 0–12, 13–21, 21–50, and greater than 50 may be an effective way to one-hot encode the age variable.

Most machine learning models work best when there is exactly one customer information per observation with its corresponding features. For example, consider a customer segmentation exercise using clustering. If the algorithm uses raw data that contains multiple lines per customer, it may treat the multiple lines as duplicates. This can introduce bias to the model for the customers with more lines. The multiple lines could indicate every change made to the customer in the system. In order to avoid this bias, the raw table can be summarized (or grouped) such that it contains exactly one row per customer and the rest of the numerical columns can provide count of the rows, totals, averages, min, max, etc. depending on the functionality of the column.

For complex data involving several thousands of features, the dimensionality of the dataset may need to be reduced before it is used by machine learning models. For example, in a machine learning application to find patients at risk of cancer, the lung x-ray images of patients need to be used to train the classifier model. As discussed earlier, due to the number of dimensions in images the raw data will need to be run through Principal Component Analysis (PCA) first to ensure one the principal components are taken into account when training the classifier model. This can improve the training process tremendously and capture the right level of information needed to train the model.

## Ensemble

The random forest algorithm seen earlier is an example of an ensemble technique where the aggregate output of decision trees is used for classification and regression problems. In general, an ensemble uses multiple machine learning techniques to get better predictive performance when compared with one machine learning technique alone.

A bucket of models is an ensemble method that uses many models such as random forest, neural networks, naive bayes, support vector machines, and decision trees to solve the same prediction problem. The algorithm along with the corresponding performance scores are tabulated to find which algorithm performs well when the problem changes. This can be a powerful tool especially for fraud detection applications where there is a continuous change to the problem environment and the best possible algorithm evaluated needs to be found to improve the true positive rate.

# Conclusion

Machine learning is divided into two main types: supervised and unsupervised. In supervised learning, labels are provided to the algorithm to train the model. Whereas, in unsupervised learning, the labels are not given to the algorithm. Unsupervised learning helps us in finding hidden structures in the dataset. Both of these types of machine learning algorithms have their own sets of performance measurement metrics. PCA can be used to reduce datasets with higher dimensions (or features) into lower dimensions without losing most of the information.

# CHAPTER 10

# Data Lakes

This chapter focuses on the role of data lakes in shaping the current IT landscape. Data lakes have become increasingly popular in the last few years and are being used widely to realize tangible business value. This chapter teaches the reader how to leverage the power of data lakes to drive their analytics requirements. It also looks at two commonly observed data lake architectures and when they are suitable for use.

## Introduction to Data Lakes

A major problem with data warehouses is that the data residing in them are preprocessed and filtered for a specific purpose and can handle structured data better than unstructured and semi-structured data. For instance, suppose financial transactions are to be analyzed in a system as part of regulations. As part of the requirements, the general ledger data containing transaction level detail may need to be reconciled with the trial balance, which includes the opening and closing balance for the month. To support this process, the general ledger and trial balance data will be loaded into a data warehouse so that it is easy for the analysts to combine general ledger and trial balance data to get the required information for reporting. The data in the warehouse may be helpful for other purposes – for example, it can be very helpful for internal auditors to use the same data to monitor for fraud.

To monitor fraud, the internal auditors may want to scan their original vendor invoice (images that are unstructured data) to match with the general ledger data that is already in the data warehouse. This would be a major undertaking for the organization to implement. Once the fraud monitoring is set up, it may or may not bring value to the company, so the Return on Investment (ROI) is not clear.

Figure 10-1 shows the setup of the proposed fraud monitoring system.



© Maris Sekar 2022
M. Sekar, *Machine Learning for Auditors*, https://doi.org/10.1007/978-1-4842-8051-5_10

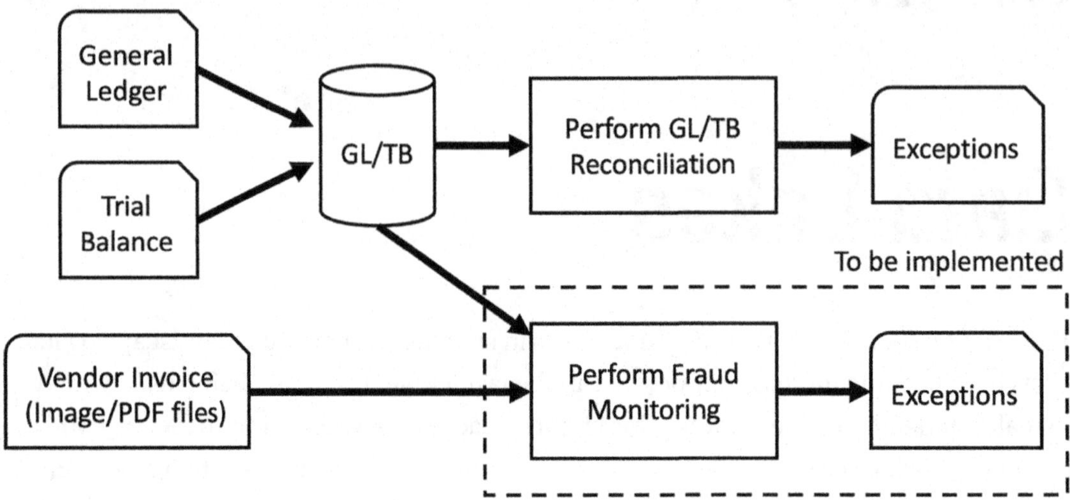


***Figure 10-1.*** *Fraud monitoring system*

The preceding setup can be used for another purpose – to monitor fraud using the original vendor invoice. If the company wants to find a way to monitor duplicate payments as part of their financial controls, this system can leverage the general ledger data as well. But this would be an undertaking costing money, people, and valuable time. What if we had a way to leverage the general ledger for all the above applications in a seamless manner and at the same time be able to combine insights derived from unstructured data such as text, image files, or video files? This is what a data lake helps achieve.

A **data lake** is a repository used to store vast amounts of raw data in its native format until it is needed. It can be used to store structured, semi-structured, or unstructured data. While a traditional data warehouse stores data in files or folders, a data lake uses a flat architecture to store data. A set of extended metadata tags are tagged to unique identifiers that are assigned to data elements in the data lake. Like a traditional data warehouse, the data lake can be queried to answer a business question at hand.

A data lake is meant to serve as a central repository for sharing organization data to different types of users such as business analysts, data scientists, data engineers, product managers, and decisionmakers. Data integrity, including data accuracy and completeness of the incoming data sources, needs to be in check in order for the data lake to be effective. Some of the benefits of using data lakes include durability, scalability, abstraction from schema, and cost-effectiveness.

## Tangible Value

A concept is only as good as its assumptions. Although the whole premise for a data lake is to centralize information and be able to leverage information and to combine with insights gained from other functions of an organization, this may not be the case in most organizations. Most organizations have a data lake for every function such as Finance, Trading, Compliance, Safety, etc. This goes against the assumption that common insights between the functions would be leveraged by each other which is where the true value of a data lake lies. If multiple data lakes exist for every function, they essentially have the same problems that data warehouses had. Multiple instances of data lakes create their own eco-systems where data is analyzed in silos. Data silos act as pockets of data warehouses that provide information for a certain purpose. This setup takes away from the flexibility in analysis that a singular data lake provides.

Figure 10-2 illustrates the problem with having multiple data lakes - one data lake for every function.

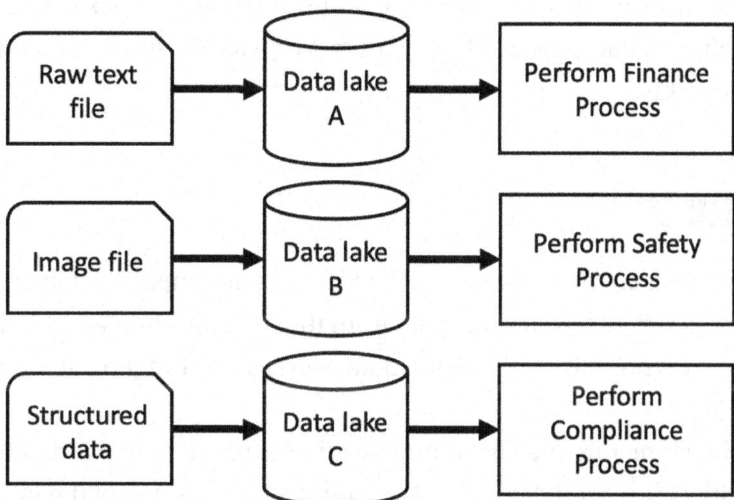


***Figure 10-2.** Data lakes by business function*

## Role as Analytics Enabler

There is one other subtle functionality that data lakes support and which was touched on in the introduction section. Different roles within the organization - data analysts, data engineers, data scientists, functional analysts, and the executive team - can all query from the same data lake to derive different insights leveraging the same underlying data. This is possible because the raw data in the data lake can be "massaged" to the purpose through a schema that sets the context for the purpose. A schema provides a structure for the data and their relationships with one another.

One of the major challenges with using data from a traditional data warehouse is there may be overdependence on other specialized analytics teams to get the insight based on their roles in the organization. Every role has their own access privileges and this can restrict the flow of information and insights. This puts a heavy emphasis on a particular analytics function and, as a result, bottlenecks start to arise based on the available resources. The analytics team becomes the bottleneck to deriving insights due to a shortage of resources.

Another major challenge is balancing governance and accessibility of information. We may restrict valuable information that could potentially provide insights because it may contain personally identifiable information. A culture of experimentation needs to be promoted so that data scientists can connect to the data to "play" with it without thinking about how to access it or rely on someone else to keep track of and extract the data on their behalf. A data lake can help with this provided a suitable architecture is chosen to support this.

# Architectures

There are two distinct architectures when it comes to implementing data lakes. In the first architecture, the data lake ingests data from the data warehouse, and in the second, the data lake collects raw data from all the data sources and forwards it to the data warehouse.

The second architecture may be a suitable architecture if empowering the organization with centralized data to power their decisions is one of the main business objectives. If data governance is one of the key priorities for the organization, then the first architecture may be the better option.

Figure 10-3 shows the first data lake architecture.

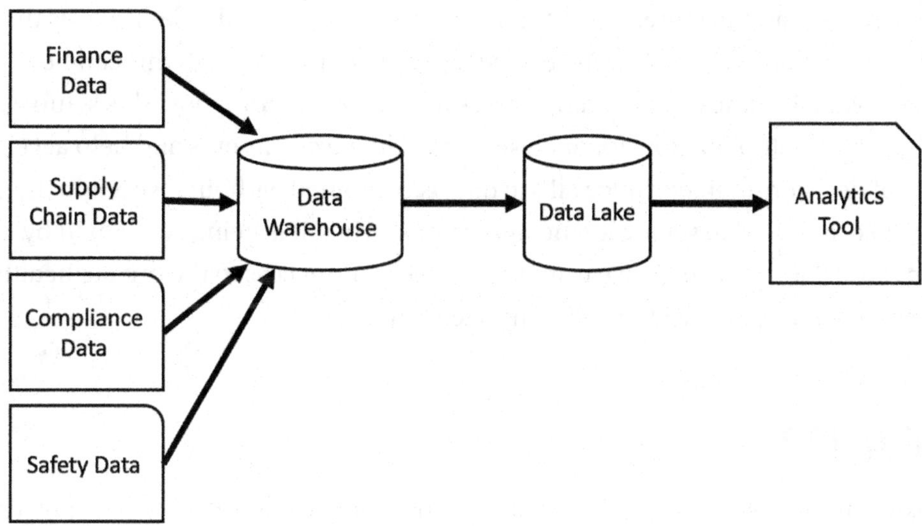


***Figure 10-3.** Data lake architecture #1 (governance oriented)*

As seen from Figure 10-3, by using the data from the business warehouse, additional controls to restrict access to sensitive information can be controlled before the data lake ingests the restricted information. Although this restricts access to certain information, it can still function as an effective analytics test environment.

Figure 10-4 shows the second data lake architecture.

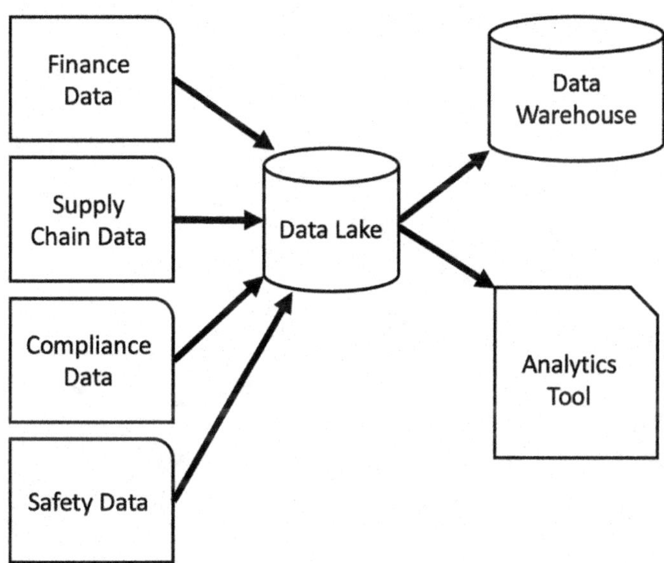


***Figure 10-4.** Data lake architecture #2 (analytics oriented)*

In the second architecture, as seen in Figure 10-3, by having the data lake as the central data repository, there is unrestricted access to data. Although this can be a nightmare when it comes to governing access to sensitive information, this setup can provide the best setup possible for analysts to run their experiments and help accelerate the pace of decision-making. Since all the data is combined without restriction in the data lake, it is hard to segregate the information when it is being consumed by an analytics tool. For instance, do we want to give everyone access to the private health information of those involved in company accidents?

# Conclusion

Data lakes can store structured, unstructured, and semi-structured data. The power of data lakes is in its ability to integrate multiple types of information from various data sources. In an organization, it is best to use a unified data lake in order to derive the best value out of them. A data lake plays a key role when it comes to enabling analytics in an organization. It serves as a sandbox environment or playground to try out new analytics use cases. There are two main types of data lake architectures: governance-oriented and analytics-oriented.

# CHAPTER 11

# Leveraging the Cloud

In order to develop machine learning (ML) applications, a development environment needs to be set up so that the algorithms can be deployed. For example, in order to detect fraud, we will need to load payments data into an environment, build the ML model to detect fraud, train the model using the data, and deploy it into production so it can be leveraged to detect fraudulent payments. In this chapter, we will look at how the major cloud providers support ML and AI development. We will first look at how a free Python development environment, particularly, Jupyter Notebooks, can be deployed on a local workstation. This will be followed by the Python development environments provided by Amazon SageMaker, Google Colab, and IBM Watson. The goal of this chapter is to introduce the reader to the high-level configurations of these environments.

## Local Workstation

Thanks to the open-source community (and the Google Brain team), the top three programming languages (arguably) for data science and ML are freely available for use and have a thriving community of support. The three most common tools are Python, R, and SQL. In this section, we will look at the general setup for Python-based integrated development environments (IDEs) for a local workstation. An IDE is a development environment that provides a one stop shop to develop applications using a programming language like Python and R. It includes a compiler or interpreter depending on the programming language being used. In the next sections, we will look at how similar setups exist for cloud deployments offered by major cloud providers.

Leveraging Python with an IDE enables data scientists to quickly create ML and AI applications using its built-in libraries. The IDEs provide a Graphical User Interface (GUI) that is easy to use and compile lines of code at a time to speed up the development process. A very popular IDE to use with Python is "Anaconda," which is a freely available



© Maris Sekar 2022

M. Sekar, *Machine Learning for Auditors*, https://doi.org/10.1007/978-1-4842-8051-5_11

software that is available for most Operating System (OS) platforms such as Microsoft Windows and Apple MacOS. Anaconda enables one to install web-based IDEs known as Jupyter Notebook and JupyterLab which are distributed by the Project Jupyter.

Figure 11-1 is a screenshot that shows the Jupyter Notebook IDE GUI in action.

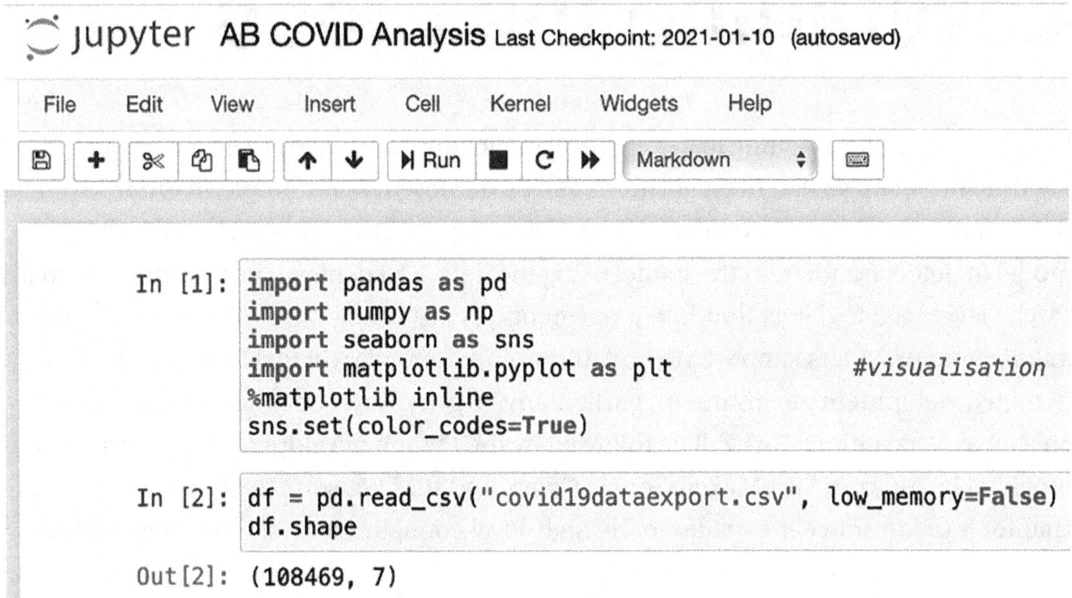


***Figure 11-1.*** *Jupyter Notebook IDE GUI*

Figure 11-2 shows another IDE also provided by Project Jupyter. Either of the IDEs can be used to develop ML and AI projects with Python code.

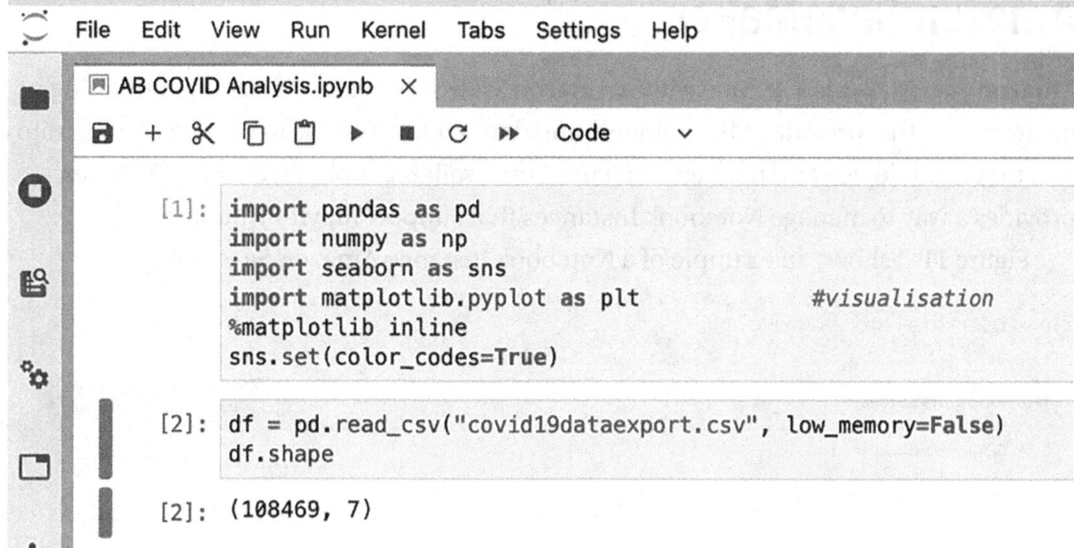


***Figure 11-2.*** *JupyterLab IDE GUI*

In this book, we will only cover at a high level what the IDE looks like and the high-level setup of some Notebooks. Coverage is at a level so that an internal auditor will be able to work with the IDEs. The assumption is that internal auditors will work with dedicated data scientists or data engineers who will help with the actual development process. For adventurous internal auditors who have previous development experience, there are many resources available to get started. The resources will be available in Appendix A.

# Cloud Computing

It is reassuring to know that the major cloud service providers support the development of AI and ML applications. Amazon SageMaker, Google Colab, and IBM Watson facilitate the creation of Jupyter notebooks as well as other tools. The same Jupyter Notebook IDE that can be installed in local workstations can be easily installed and often provided with native support (no installation required) with the above-mentioned cloud environments. This makes it easy for small organizations looking to get started with no or minimal infrastructure to implement ML and AI using their organizational data. Many organizations are also already on the cloud. This makes it even easier to add support with the other available tools based on the existing environment. In the next sections, we will explore how each of the cloud providers enable support and will introduce the reader to the basic setup.

# Amazon SageMaker

Amazon SageMaker is a service within Amazon Web Services (AWS) cloud computing environment that provides ML engineers and data scientists tools to prepare, build, train and tune, and deploy and manage. As part of the available tools, Amazon SageMaker provides a way to manage Notebook Instances that support Jupyter Notebooks.

Figure 11-3 shows an example of a Notebook Instance Amazon SageMaker.

Amazon SageMaker > Notebook instances

**Notebook instances** Actions ▼ Create notebook instance

Search notebook instances

| | Name | Instance | Creation time | Status | Actions |
|---|---|---|---|---|---|
| ○ | SageMakerCourse | ml.t2.medium | Jan 24, 2021 04:06 UTC | ⊖Stopped | Start |

***Figure 11-3.** SageMaker Notebook instance*

An instance needs to be created and started. Once the instance is started, you can run Jupyter Notebooks or JupyterLab on this instance.

Figure 11-4 shows an example Jupyter Notebook within a managed instance on Amazon SageMaker.

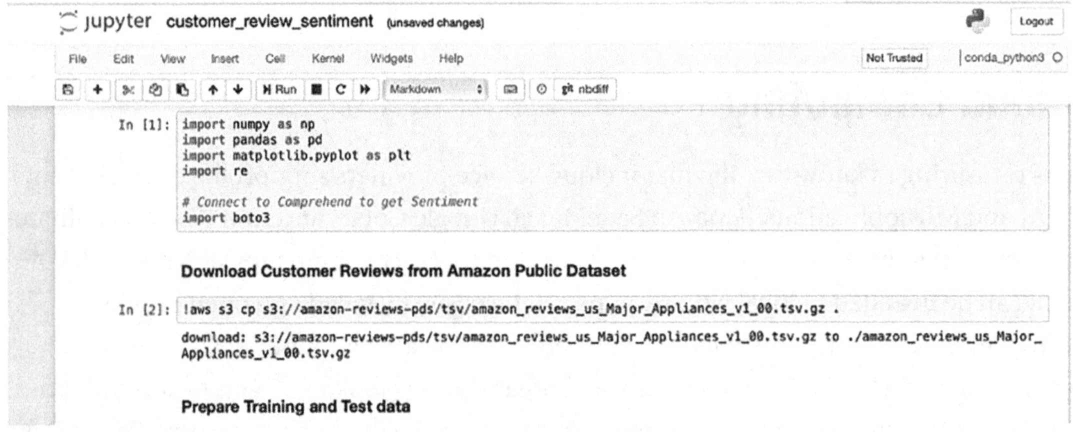


***Figure 11-4.** Jupyter Notebook within Amazon SageMaker*

The instance can load data from and export data out of Amazon S3 buckets. Amazon S3 buckets is the cloud storage system available on Amazon. S3 stands for Simple Storage Service. Boto is a client allowing the consumption of an Amazon SageMaker service. Boto provides APIs that can be used to create and manage Amazon SageMaker instances.

## Google Colab

Google provides a similar Python development environment which is deployed on the Google Cloud.

Figure 11-5 shows a Notebook in the Google Colab environment.

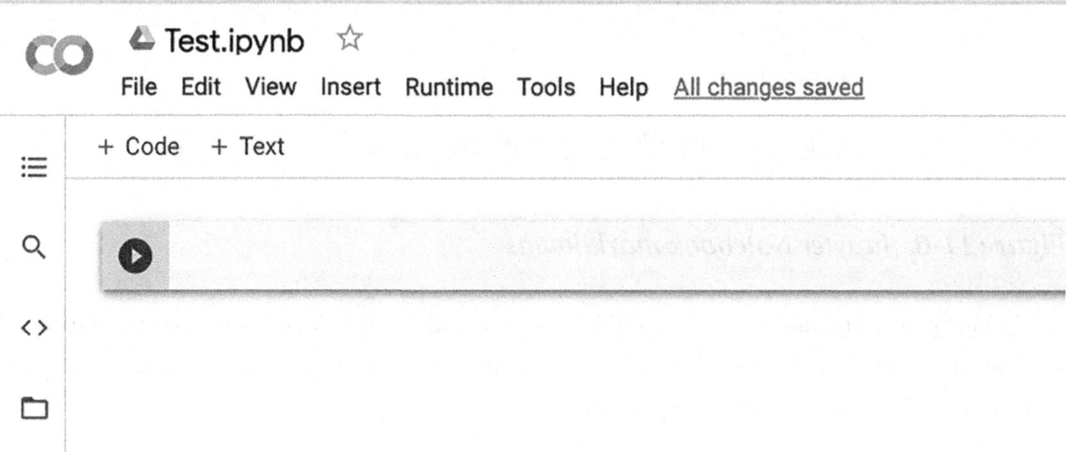


***Figure 11-5.*** *Google CoLab Notebook*

Google Colab has the same features as Jupyter Notebook, but it provides an easier way to represent markdowns. Markdowns is an interesting feature in Notebooks that helps to keep your notebook tidy and combine non-code text with code within your Notebook.

In Jupyter Notebooks, you can specify if a cell is a Markdown or Python code as shown in Figure 11-6.

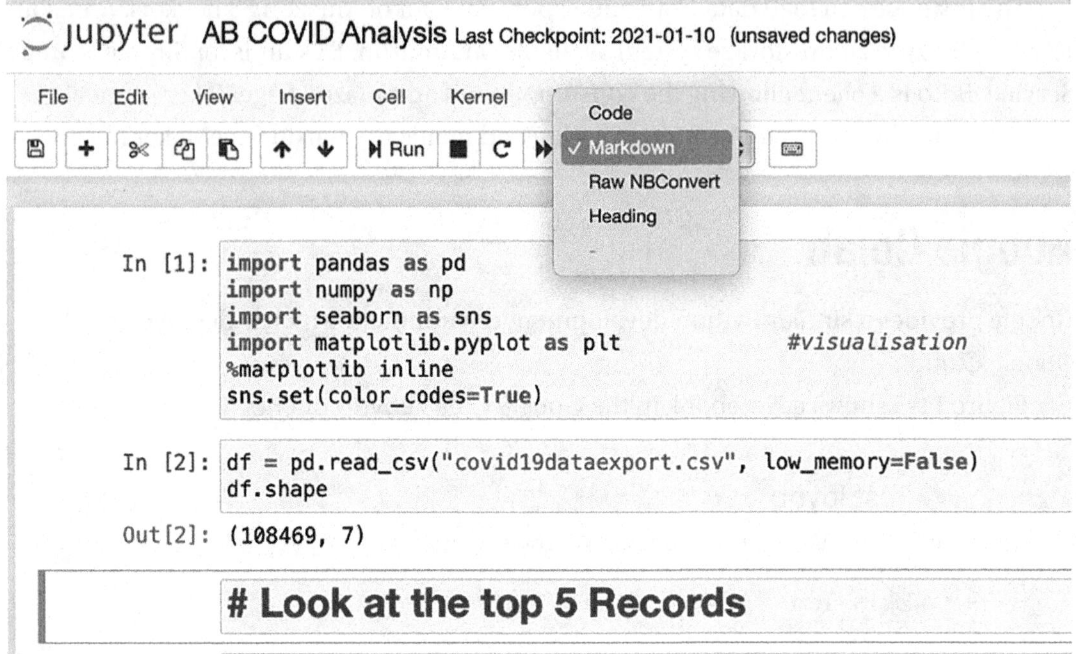


***Figure 11-6.** Jupyter Notebook markdowns*

In Google Colab, as shown in Figure 11-7, you can click "+ Text" button on the top toolbar to create a markdown. Colab also provides additional tools for formatting, so you do not need to remember the formatting syntax.

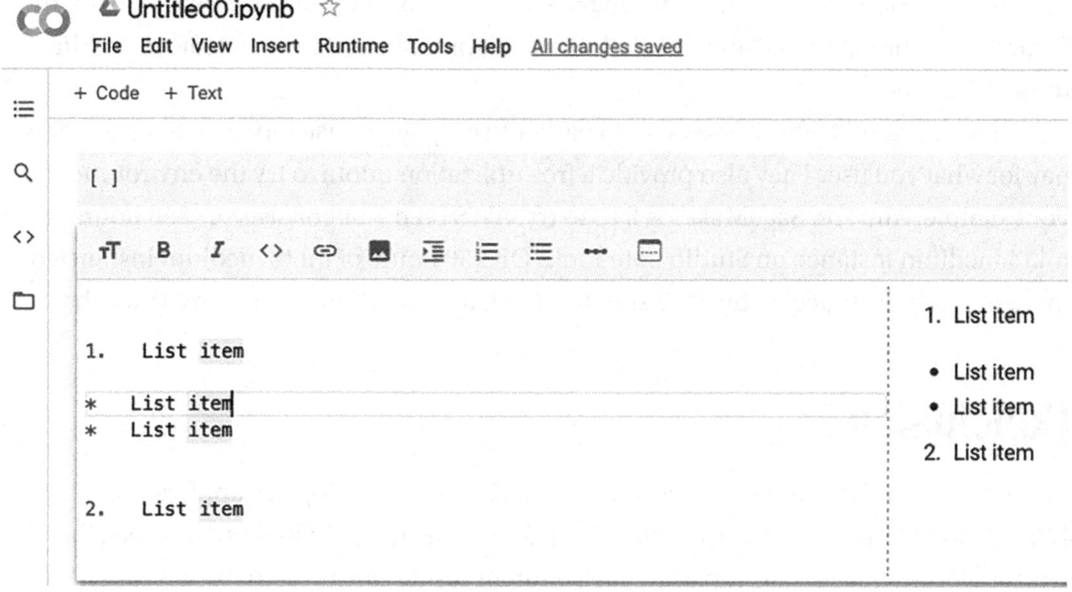


*Figure 11-7. Google Colab markdowns*

# IBM Watson

IBM Watson Studio is a service provided by IBM Cloud to build, run, and manage AI models and optimize decisions. IBM Watson also provides a Python development environment.

Figure 11-8 shows a Python Notebook within the IBM Watson Studio.

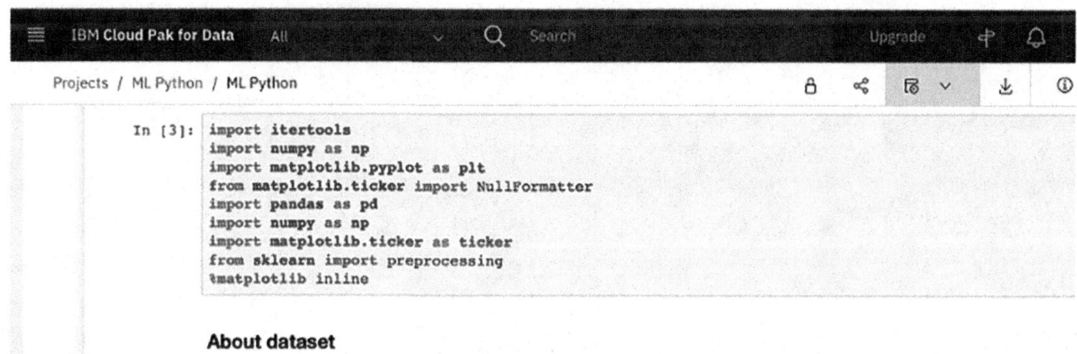


*Figure 11-8. IBM Watson Notebook*

There are other features in Amazon SageMaker, Google Colab, and IBM Watson. In Figures 11-3 through 11-8, only the Python development environment feature within these solutions.

All of the cloud technologies shown earlier use a pay per use model, where you only pay for what you use. They also provide a free utilization quota to try the environments. For example, Amazon SageMaker is free to try. AWS Free Tier consists of "250 hours of ml.t3.medium instance on Studio notebooks OR 250 hours of ml.t2 medium instance or ml.t3.medium instance on on-demand notebook instances" for the first two months.

# Conclusion

There are many development environments available to develop ML applications. Jupyter Notebooks is widely used to develop ML applications in Python. In a local workstation setup, the development environment resides locally and the analysis is not accessible for use from other workstations. In a cloud setup, the development environment can be used by any workstation and the analysis can be shared with other workstations. The three most commonly used for developing ML applications are Amazon SageMaker, Google Colab, and IBM Watson.

# CHAPTER 12

# SCADA and Operational Technology

Operational Technology is a generalized term used to denote any device used to manage operations at an industrial scale. The chapter covers the Fourth Industrial Revolution and how it provides a way to combine artificial intelligence (AI) and the Internet of Things to solve problems in the field of SCADA auditing. The chapter also covers the six main areas of the SCADA security framework. And finally, how AI can be applied to three of the areas: Data and application security, System assurance, and Monitoring controls.

## Fourth Industrial Revolution

According to Klaus Schwab, Founder and Executive Chairman of the World Economic Forum, the Fourth Industrial Revolution is building on the Third Industrial Revolution of using information technology and electronics to automate production. Internet of Things (IoT) is one technology that is transforming the production, management, and governance landscape. Other technologies of the Fourth Industrial Revolution include artificial intelligence, Robotics, and Quantum Computing.

Figure 12-1 shows the relationship between OT, ICS, and SCADA.



© Maris Sekar 2022

M. Sekar, *Machine Learning for Auditors*, https://doi.org/10.1007/978-1-4842-8051-5_12

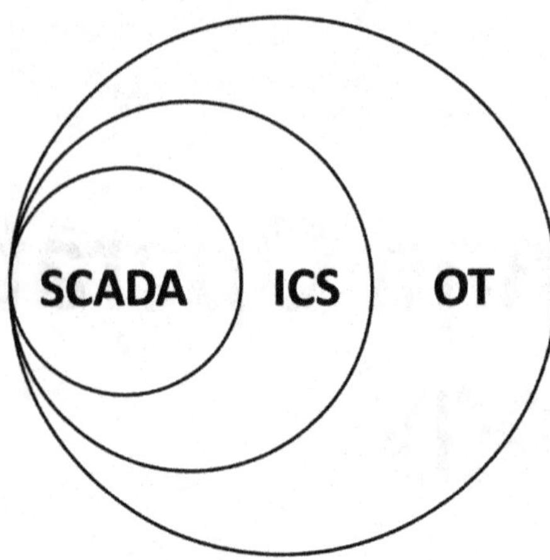


***Figure 12-1.*** *Relationship between OT, ICS, and SCADA*

OT/ICS/SCADA systems have been used for some time and maybe taken to be part of the Third Industrial Revolution. OT stands for Operational Technology and represents the use of computing systems to manage industrial operations in general. ICS stands for Industrial Control Systems. It is a subset of OT used to monitor and control industrial processes such as pipeline monitoring and control for the oil and gas industry. SCADA stands for Supervisory Control and Data Acquisition and typically provides graphical user interfaces (GUI) to observe and control systems. SCADA is a subset of ICS.

IoT is primarily concerned with the usage of computing devices to support system-to-system communication and analysis of data to improve the productivity of a specific task. IoT devices function as autonomous systems that support some intelligent processes for processing data.

The audit of SCADA systems poses unique challenges such as accessibility and maintenance of systems. Most of the SCADA systems support the control and acquisition of data from devices that are in remote locations. The IT Auditor would typically travel to these locations to audit the systems. Due to budget and availability constraints, only a small subset of sites can be audited for a particular year. Another major challenge with auditing SCADA systems is that the systems are often not connected to the corporate network due to cybersecurity reasons. The SCADA network may need to be segregated from the corporate network so that attacks on the SCADA network do not affect the corporate network and vice versa. Due to the systems being disconnected from the corporate network, the patching of SCADA systems' Operating System and Anti-virus can be tough to overcome. These barriers can lead to compromised SCADA systems that are vulnerable to cyber-attacks. This chapter proposes the use of AI and IoT to audit SCADA systems.

## SCADA Auditing

According to an article published by Samir Malaviya for the Information Systems Audit and Control Association (ISACA) Journal titled "SCADA Cybersecurity Framework," the SCADA security framework comprises six areas.

The six areas are

- **Governance, risk, and compliance administrative controls** - These controls pertain to the policies, standards, rules, and exception management. The controls are derived from the Governance, Risk, and Compliance (GRC) framework and are non-technical in nature.
- **SCADA controls** - SCADA poses its own unique challenges such as the accessibility and patch management problems discussed earlier. They also need to maintain segregation of networks. These requirements are captured as part of the SCADA controls.
- **Data and application security** - Third-party vendors often develop SCADA software. It is essential that the SCADA software development lifecycle be standardized to meet software development guidelines. Data from SCADA systems should be readily available and accurate, and confidential. Data security helps to mitigate risks related to SCADA data.
- **System assurance** - The SCADA system as a whole may need to be available. Business Continuity Planning (BCP) and Disaster Recovery Plans (DRP) will need to be put in place to ensure the business continues to operate when there is an unexpected interruption.
- **Monitoring controls** - SCADA incidents will need to be handled in an efficient and standardized manner. In addition, the SCADA systems will need to be continuously monitored for threats to ensure that suspicious activity is followed up on.
- **Third-party controls** - These controls pertain to controls used to manage vendors who have access to the SCADA systems. Additional controls may need to be implemented to ensure vendors do not abuse their administrative access to the system either intentionally or unknowingly. Poor third-party controls can result in unauthorized access to data or unavailability of data.

The following section gives details on applying AI to Data and application security, System assurance, and Monitoring controls.

## Applying AI to SCADA Auditing

Predictive maintenance is a concept that is starting to get traction in the field of operations management. Predictive maintenance allows us to monitor the system state with the help of data collected from the machines to predict system failures in advance so that unscheduled downtime can be avoided. ML can be used in predictive maintenance to prevent these system failures. ML in predictive maintenance can optimize and improve the predictive power of system failure so that the flagged system can be repaired before it needs to be completely shut down for maintenance. For example, data from gas compressors in remote field locations can be collected, and compressors that fail over time can be captured within the data along with the operational data retrieved from the compressor's measurement systems. Using machine learning, a system can predict what series of events or characteristics cause the compressors to fail. The identified systems can be serviced well ahead of the intended failure timeline to avoid costly repairs or shutdowns later on.

The preceding concept can similarly be used to identify SCADA workstations that are due for operating system and antivirus upgrades. By continuously monitoring the workstations due for service (with the help of machine learning), a field operator can be notified when workstations are up for service when they are in the area. The field operator can then perform the necessary patches and updates.

Machine learning models can be developed to detect anomalies with the data collected from the workstations. For example, other machines that are typically connected with the SCADA network can be learned by the machine learning model. When an unidentified device is connected to the network, the Cybersecurity or IT team can be notified for further action. The machine learning model can reside on the SCADA network and monitor the SCADA workstations connected to the network. It is important for the ML model to reside on the same SCADA network to prevent security breaches on the corporate network. The corporate network is more susceptible to cyberattacks due to its internet connectivity when compared to the SCADA network. SCADA networks are typically not connected to the internet.

Another use case for anomaly detection in the field of SCADA auditing is to identify outliers for reviewing user access management. A matrix can be created to plot clusters of users' roles along with the access they hold. The outlier users can then be identified and followed up on to ensure these are authorized users. This can save a lot of time and effort on the part of the IT Auditor when verifying user access on the SCADA network. It can be automated so that the flagged users are continuously reviewed by the IT or Cybersecurity team.

Figure 12-2 shows a sample of the user access management review using clustering.

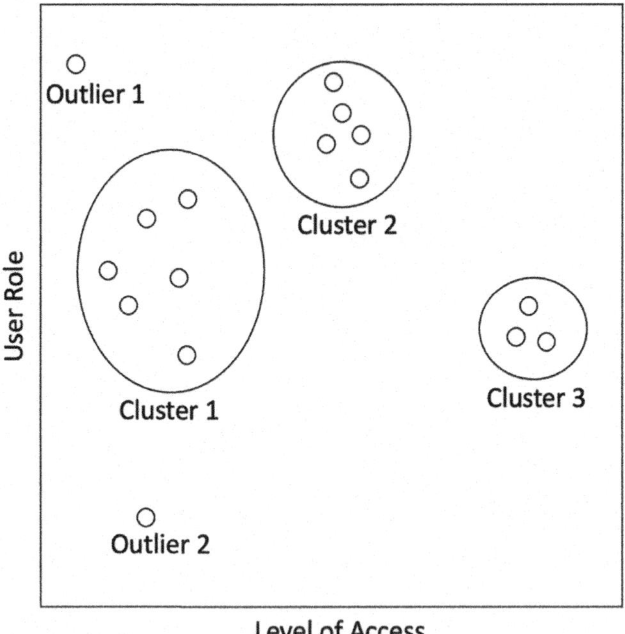


***Figure 12-2.*** *Anomaly detection for user access management*

User roles with similar levels of access form groups of identifiable clusters. Outlier 1 and Outlier 2 are roles that appear to be anomalies in the data and warrant a closer look at their setup.

# Conclusion

The fourth industrial revolution presents new opportunities for Operational Technology such as SCADA. The application of AI to SCADA can provide many use cases for SCADA Auditing. The two main types of ML application use cases are predictive maintenance and anomaly detection for user access management.

# PART III

# Storytelling

CHAPTER 13

# What Is Storytelling?

Data Storytelling creates the best business outcomes by telling effective stories about the available data. It builds on basic elements of data analytics and combines it with aspects of creative storytelling to make an impactful data story. This chapter goes through the major elements of data storytelling. The common pitfalls when it comes to crafting data stories are also discussed here. Finally, keeping the audience engaged and focused is an important aspect of storytelling. This is covered at the end of the chapter.

## Data Storytelling

Data Storytelling is an important part of effective data analytics. Let us demonstrate this using an example. Visualizations are one of the fundamental blocks when it comes to data analytics. Suppose you were performing a privileged access audit in your company. One of the objectives of the audit is to see if there are privileged users in the audited system. You can perform the anomaly detection algorithm for user access managed as discussed in the previous chapter.

Figure 13-1 shows the result of running the anomaly detection for user access management.



© Maris Sekar 2022
M. Sekar, *Machine Learning for Auditors*, https://doi.org/10.1007/978-1-4842-8051-5_13

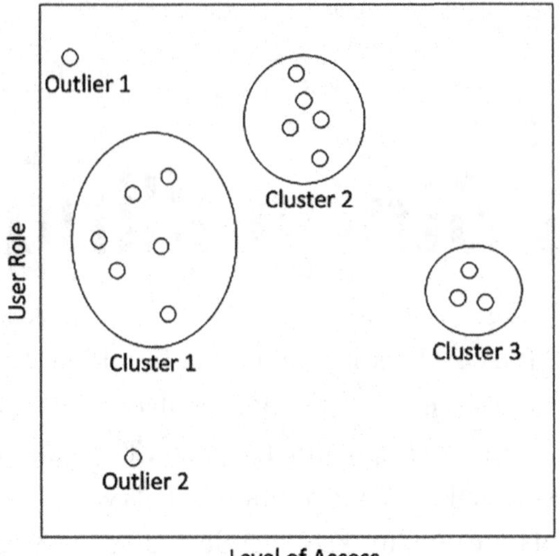


***Figure 13-1.*** *Anomaly detection for user access management*

Figure 13-1 demonstrates the outliers clearly. But what if we show the following visualization instead (as shown in Figure 13-2). Figure 13-2 includes colors (black and white) to highlight the privileged users in the system. It has an action title – a title that clearly conveys the message of the visualization.

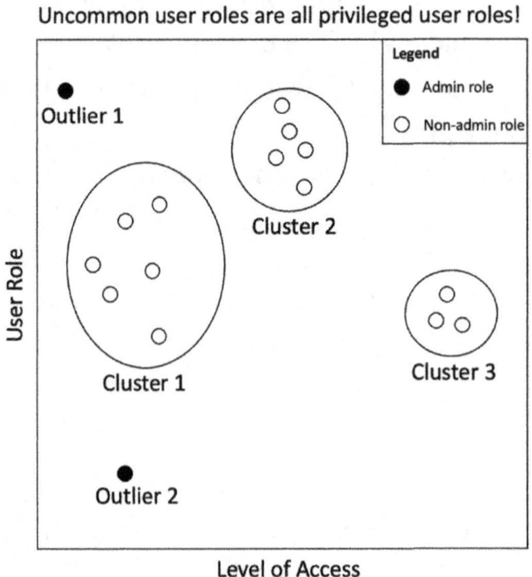


***Figure 13-2.*** *Anomaly detection for user access management*

Clearly, the second diagram conveys a better story and contains some of the best practices of telling a good data story.

Let us look deeper into the major elements of data storytelling as illustrated in Figure 13-3. Many of the elements agree with creative storytelling with a major emphasis on setting the context of the story.

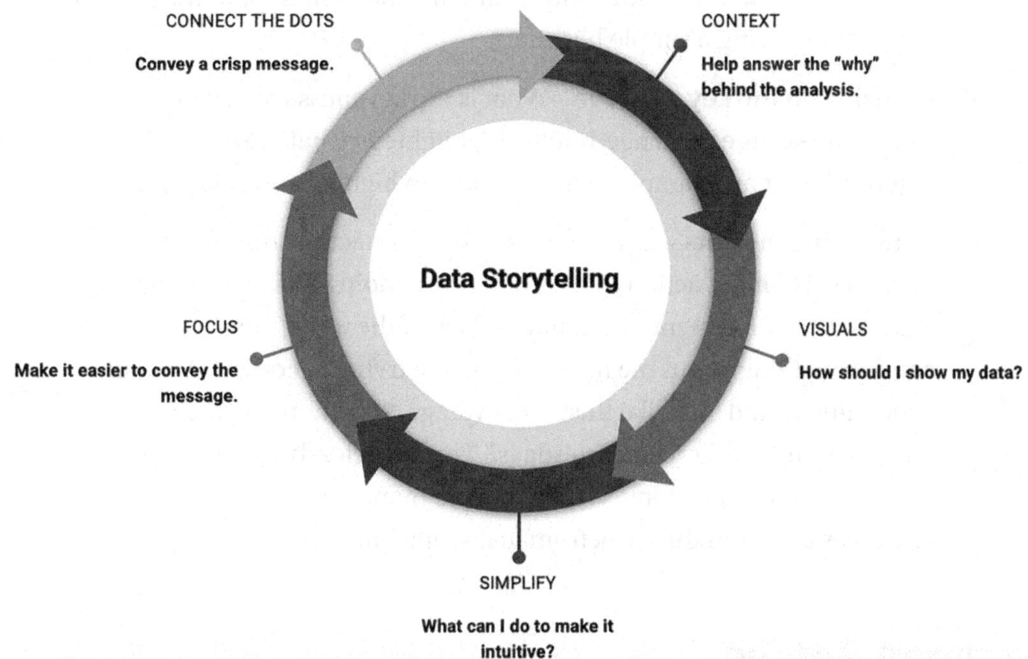


***Figure 13-3.*** *Elements of data storytelling*

The five elements of creating effective stories are

- **Setting context** - Define the setting or the environment. What environment did the data come from? How does it relate to the business problem at hand? Does it help answer why a particular data story needs to be told?
- **Visualizing data** - What are the visuals I need to use to display the data? Would a bar chart suffice or do I need a scatter plot (or some other visualization)? Data visualization helps to see the data in an understandable manner.

- **Simplifying the content** - Often there is a general belief that the more visuals we add or the more dimensions (data fields) there are, the more value the analysis brings forth. It is typically better to use a few simple visualizations to communicate the meaning of the data to your audience. If a bar chart can convey the same information you are trying to show with a multi-dimensional scatter plot, we recommend using a simple bar chart.
- **Focusing on the key message** - What is the key message of your story? The key message needs to be highlighted visually. See Figure 13-3 for an example of using colors to highlight your message.
- **Connecting the dots** - Lastly, you need to connect the dots for your audience. Defining action titles is one way of doing this. For example, a title such as "Payment frequency vs. Day of the week" may not convey anything about the message you are trying to convey. A better action title would look like this: "On average, 20% more payments are made on Mondays and Tuesdays." This title clearly explains the message you want to convey and a people manager can plan more resources earlier in the week from just seeing the title.

# Common Pitfalls

There are many common pitfalls associated with data storytelling. We will cover the most common ones in this section.

# Misleading Graphs

This is a common problem seen in many visualizations. This can be exploited on purpose or can be unintentionally introduced into the visualization.

In the first set of visuals shown in Figure 13-4, all of the categories have exactly four items. This demonstrates that the visuals can mean different things if they are not properly set up.

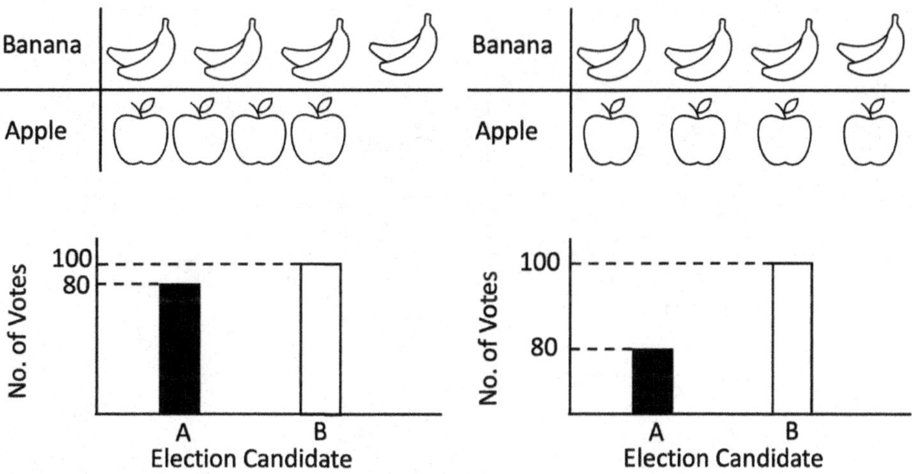


***Figure 13-4.** Examples of misleading graphs*

In the second set of visuals, the two graphs represent the same information, but the left one shows them being closer when compared to the graph on the right. Sometimes this is intentional. For example, in an election, to make it look like it was a close election the left graph would be shown. The right graph may be shown to indicate that candidate B won by a big difference.

## Anscombe's Quartet

Anscombe's Quartet was created by a statistician named Anscombe to illustrate the danger of using summary statistics to understand data. To fully understand the nature of the data, its distribution must be seen to get a more accurate picture. Averages are a common way of simplifying the data and if you use them, you need to keep in mind that you will be losing some information by summarizing it.

Figure 13-5 illustrates the Anscombe's Quartet in the form of four scatter plots.

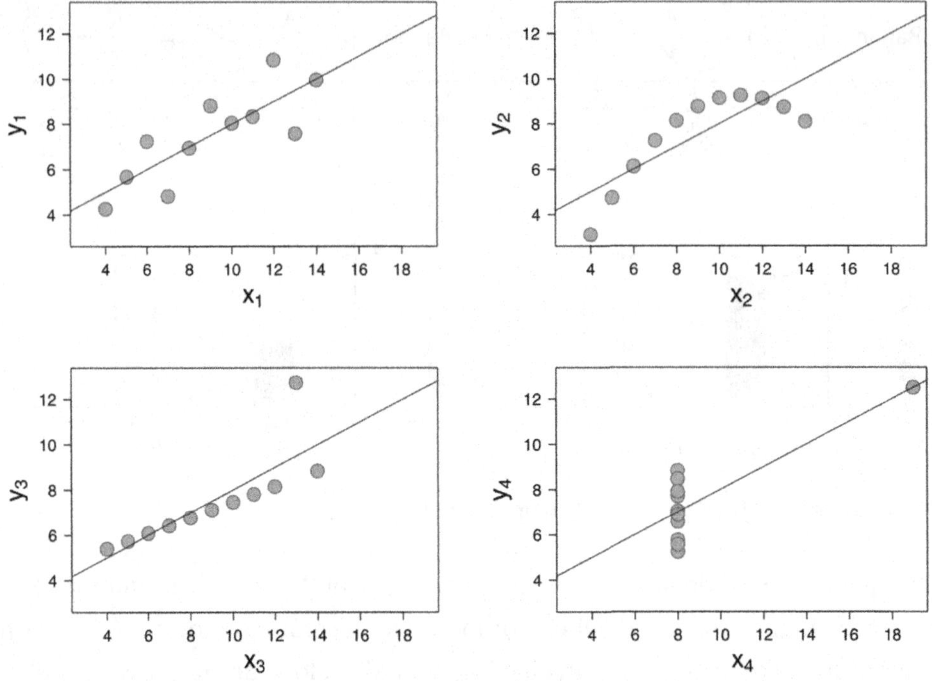


***Figure 13-5.** Anscombe's Quartet graphs*

Observe the four graphs shown in Figure 13-5. All four have three similar properties. They have the same mean, variance, and correlation.

# Engaging the Audience

One of the easiest ways to engage the audience is to be aware of the Z-Pattern. The Z-Pattern was found through researching how people normally read diagrams. As depicted in Figure 13-6, the general finding was that people normally start from the top left corner of the diagram and then move to the top right corner of the screen, followed by the bottom left and finally, the bottom right corner of the screen. The focus gravitates from the top left corner of the screen towards the bottom right corner of the screen. What this means is that you want to put the most important visuals on the top left and then on the top right followed by bottom left and lastly, bottom right.

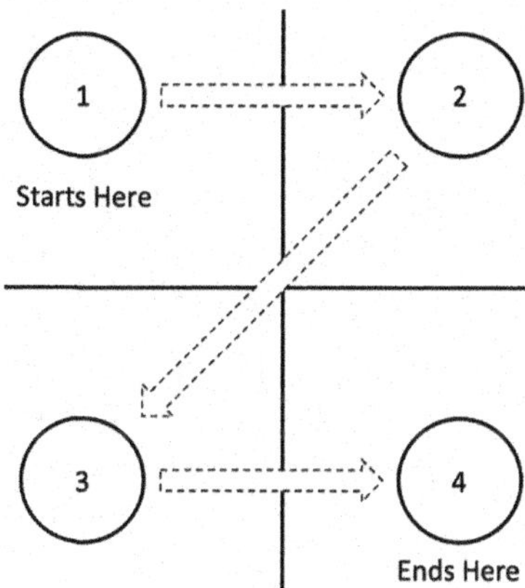


***Figure 13-6.*** *The Z-Pattern Layout*

Gestalt Design Principles can also be leveraged to build visuals that convey the message clearly and to maintain audience focus. Gestalt principles are design principles that have predominantly been used in User Interface (UX) design. They can also be used to develop data analytics dashboards and visuals.

# Conclusion

Data storytelling can be leveraged to share impactful findings in the audit. The main elements of data storytelling are: Context, Visuals, Simplify, Focus, and Connect the Dots. Two of the common pitfalls of storytelling include misleading graphs and Anscombe's Quartet. It is important to engage the audience using available tools and techniques such as Z-Pattern and Gestalt Design principles. Paying attention to these techniques can help make an effective presentation.

# CHAPTER 14

# Why Storytelling?

This chapter builds on the previous chapter by focusing on why data storytelling works. We will look at a practical IT Audit example of how storytelling can be used to enhance insight generation. We will also cover the general guidelines of good storytelling. Finally, a summary of what to do to craft good data stories as well as what not to do when creating them is listed for reference.

## Why Does It Work?

Many times, the person generating the insights is different from the person that can take action based on the insights. Insights need to be interpreted easily from visuals with minimal guidance. How that visual is communicated to the person taking the action is as important as generating the insights themselves. As stated in the previous chapter, the following are the major elements of data storytelling:

- Context
- Visuals
- Simplification
- Focus
- Connecting the dots

These key elements enable us to tell a concise and impactful data story that can be easily interpreted and actioned upon. Among these, Context is the most crucial component. It helps address two key questions – why is the insight performed and how can it be useful?

Sometimes, the algorithms being used to derive insights may be complex or the visuals may not be enough. In these cases, it is important to break down the insights into simpler and easily digestible content so that it can be actioned upon. Consider the



© Maris Sekar 2022
M. Sekar, *Machine Learning for Auditors*, https://doi.org/10.1007/978-1-4842-8051-5_14

IT audit scenario of analyzing user activity logs. From an IT audit point of view, user logs can be used to determine if only authorized personnel have access to Personally Identifiable Information (PII). The logs can also be used to observe if the system, processes, or appropriate people resources are utilized. For example, the company policy may require that a functional owner and the application owner need to approve access to people requesting access to the system. There can potentially be several insights derived from the analysis supported by the corresponding visuals. Here are some insights that can be derived from the analysis:

- Only people within the HR team have access to PII.
- All people requesting access have a functional approver from the appropriate team and the application owner.
- Some terminated users still have access to the system.
- System seems to be utilized more during the non-summer months.

Although the above insights are interesting, they may not all be reported to the same team. The insights need to be forwarded to the appropriate people who are responsible to mitigate the risks arising from them, if applicable. Therefore, the context of the audit and its scope is important. Insights that are not within the current audit scope may still need to be actioned by the appropriate party based on the criticality of the finding.

Another reason good stories are often effective is because they help to convey a key message in a non-technical manner and propose to connect with the audience.

# General Guidelines of Good Storytelling

The general guidelines of a good story are covered in this section. Dashboards are a common way to present findings to end-users. A proper dashboard's layout is logical and easy to read.

# General Dashboard Layout

The most important Key Performance Indicators (KPIs) are located at the top left and more KPIs or summaries at the top right corner of the screen. In the middle, you would put in more summary graphs and lastly, at the bottom third of the page, detailed data (mostly a data table) is provided for the user. Please refer to the dashboard example given above for an illustration of this.

Figure 14-1 demonstrates the general layout for a dashboard.

In Figure 14-1, notice how the top left corner of the page with the safety traffic cones shows the breakdown of the top events produced by the driver. This is the most important information for the user. The gauge on the top right is the second most important information provided by the dashboard.

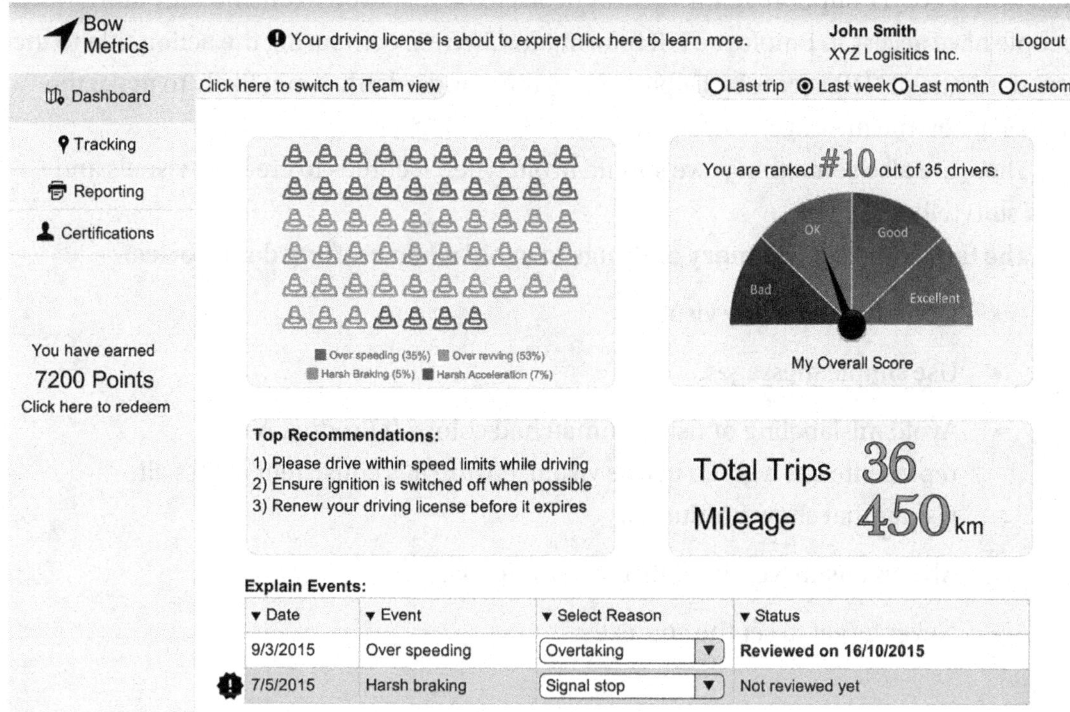


***Figure 14-1.** General dashboard layout*

**Here are the top recommendations for optimal storytelling:**

- Spend a great amount of time understanding the business and the available data.
- Data quality needs to be verified before data is used.
- Remember to set the context for your data visualizations.

- Put the most important message/visual at the top left corner of the screen.
- Keep your visuals simple and clean.
- Use action titles to lead your audience to your key message.

An example usage of the action titles would be to replace the standard "y-axis by x-axis" title with the key message the visual is trying to convey. One example of an action title would be: "IT Support Team's access needs to be reviewed to ensure only authorized people have access to Employee HR information." When comparing the action title to the standard title of "HR access by Department," the standard title is not likely to make the reader act on the message.

There are also some things we should avoid when it comes to creating visuals and data storytelling.

**The following is a summary of things to avoid when crafting data stories:**

- Don't use too many visuals.
- Use simple messages.
- Avoid mislabeling or using unmatched colors. If Product A is represented as a green in one visual, ensure it is consistent across all visuals that show Product A.
- Always ensure you identify the axes clearly.
- Never forget to set the context!

# Conclusion

The key elements of data storytelling can help the auditor tell concise and impactful data stories that can be easily interpreted and actioned upon. A good dashboard layout starts with the KPIs at the top and are followed by the summary graphs in the middle of the page. The details of the data, in the form of a table is included at the bottom of the dashboard.

CHAPTER 15

# When to Use Storytelling?

Great stories help to inspire an intended action. Audit Reporting provides many avenues when it comes to using stories. Communicating audit recommendations, conveying the impact of the findings, and summarizing the key findings are a few ways storytelling can be leveraged in the audit process. It is better to focus on two to three key messages when it comes to stories, and the business value derived from the stories should determine if storytelling is the most effective tool.

## Use Stories to Inspire or Motivate an Action

Stories can be effective in inspiring or motivating the responsible party to take action. When applying artificial intelligence (AI) to auditing, it is easy to get caught up in the details of the techniques, and the criticality of a recommendation may not be clearly communicated. Storytelling can help set the context and focus on the key message given the preceding complexities and help guide the decisionmaker on the correct course of action. Stories can be used to reiterate the points after sharing the results to make it easier for the responsible party to remember.

Here is an analogy to drive home the point. Think of raw data as building blocks of different sizes, colors, and shapes that are scattered across the floor. Processed data is similar to grouping the blocks into their respective sizes, colors, and shapes so that it can be used to build structures. In this case, data analytics can be thought of as the organization of the blocks into specific objects like windows, doors, walls, and foundation.

Storytelling is putting together the objects consisting of the building blocks into a finished house. The completed house shows how the foundation goes at the bottom of the house followed by walls, doors, and windows. One can visualize a character (made of blocks) using the door to get into the house and sitting down in the house in front of a window. Maybe the door is too small for the character to get into the door, and it may need to be adjusted. This can be visually shown to the observer and adjusted.



© Maris Sekar 2022
M. Sekar, *Machine Learning for Auditors*, https://doi.org/10.1007/978-1-4842-8051-5_15

In the field of auditing, recommendations are produced in the form of a formal audit report to stakeholders. Stories can be constructed based on the observations and their respective recommendations. The action needed by the responsible party can be clearly communicated and the expected end result can be described in the story. The stories can also be used as a way to describe the machine learning (ML) algorithms used in audit testing in layman terms so that it is clear why a particular algorithm was used and how the algorithms were leveraged. The detailed specifics of the parameters of the algorithms can be attached with supporting documentation for re-performance at a later time.

The medium of communication for stories is also important. Stories can be shared in audit reports in writing, in meetings, or memos. Among these mediums, in-person meetings can be most effective in telling stories to make the most impact and to answer any follow-up questions others may have. Audit follow-up meetings or close-out meetings can be used to share stories. Stories can be used to set the context at the beginning of the engagement, and they can also be used to summarize takeaways to nail home the key message at the end.

## When Can We Use Storytelling?

Let us revisit the audit Reporting process for data science and ML Projects.

Figure 15-1 shows the different steps involved in the audit reporting process.

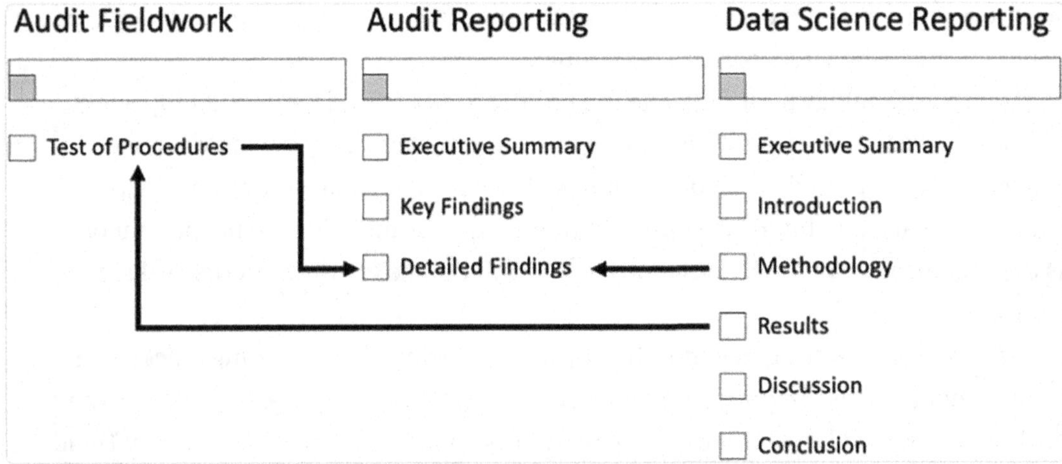


***Figure 15-1.*** *Audit Reporting process*

During the fieldwork phase of the audit, the audit program is tested and the findings are summarized in the Audit Reporting phase. Executive Summary, Key Findings, and Detailed Findings are part of the Audit Reporting phase. As an extension of the audit reporting, we must define the sections for data science applications when conducting the audit. The Executive Summary, Introduction, Methodology, Results, Discussion, and Conclusion are standard sections to include for data science projects. The relationship of the various sections within the phases are illustrated in the diagram. In particular, the methodology and results from the data science reporting phase need to be included as part of the audit reporting phase in order to provide the right level of information for another auditor to be able to re-perform the audit.

Storytelling can be used to share Key Findings as part of the Audit Reporting process and for showcasing the main results from the testing process. These are the best candidates because we need to encourage stakeholders to make decisions based on what we find through the audit testing process. The other steps involve internal communication of the audit process and do not specifically involve the sharing of results. This is a key distinction, and it is important to note that sharing of results with others in the audit process is where we can make the most impact to encourage others to make better decisions.

Naturally, the Results, Discussion, and Conclusion sections of the data science reporting phase will contain some form of storytelling in order to better present the results and come to a valid conclusion. This is expected for data science projects in general.

## Less Is More

Less is more when it comes to creating an impactful message. Is it better to state three key messages or present a story for each of the findings? Should we group the findings into high-level sections and focus on 2–3 key findings? These are questions that may arise when performing audits and trying to use stories to present them. It is questions such as these that point to the fact that applying stories to provoke the right kind of action is as much an art as it is a science.

In simple terms, it is better to focus on 2–3 key messages to tell stories. Some audit reports may have 20 findings distributed across five main sections. For example, for a cybersecurity penetration testing audit, 20 findings can be rolled into distinct sections such as social engineering, internal penetration testing, external penetration testing, vulnerabilities, and malicious software. In this case, it may be better to do a story for each of the five sections that highlight the most critical finding in each of the sections.

The following are sample key messages for each of the sections:

- **Social Engineering.** External party was able to gain password through a simulated phishing attack.
- **Internal Penetration Testing.** An internal workstation was compromised due to incorrect configuration.
- **External Penetration Testing.** Internet-facing server has default passwords.
- **Vulnerabilities.** Ten workstations do not have the most updated versions of the Operating System, leaving them exposed to known security vulnerabilities.
- **Malicious Software.** Antivirus is not running for five workstations.

Stories can be visualized, and summary statistics can be provided when looking at numbers. For example, 5 out of 1000 workstations may be a lower priority finding than 5 out of 25 workstations. Remember that stories are there to help us provoke the initial interest and take the right action. Hence, it is better to simplify and focus on the key message. When the responsible person is about to take action and requests more details on the findings, we can provide additional details and ensure all findings are resolved.

# Conclusion

Creating stories is time-consuming and may not be applicable for all audit projects. Ultimately, the business value derived from crafting stories and their impact should dictate if stories should be used. It all comes down to if the stories are effective in communicating the audit recommendations and the required actions are conveyed to the responsible person clearly.

# CHAPTER 16

# Types of Visualizations

This chapter explores the different types of visualizations and a brief overview of when they can be utilized. Some basic visualizations and a few advanced visualizations are discussed here. Other visuals can convey information, but the ones specified here cover the most commonly used and supported in Python.

## Basic Visuals

The top 10 of the most used visualizations are reviewed in this section. Although the visuals presented here are considered fundamental, they may be adequate to represent about 80% of the information and data graphically.

Figure 16-1 illustrates a scatter plot. In this plot, the CPU Utilization is plotted against the number of users. The CPU Utilization increases with the number of users. This suggests that they have a linear relationship with one another.

Scatter plots are flexible charts that show the relationship between two numerical variables. Additional dimensions can be displayed with colors and shapes of the data points being displayed. Figure 16-1 is a scatter plot showing a positive linear relationship between the number of users and their CPU utilization, that is, as the number of users increases; the CPU utilization goes up.



© Maris Sekar 2022

M. Sekar, *Machine Learning for Auditors*, https://doi.org/10.1007/978-1-4842-8051-5_16

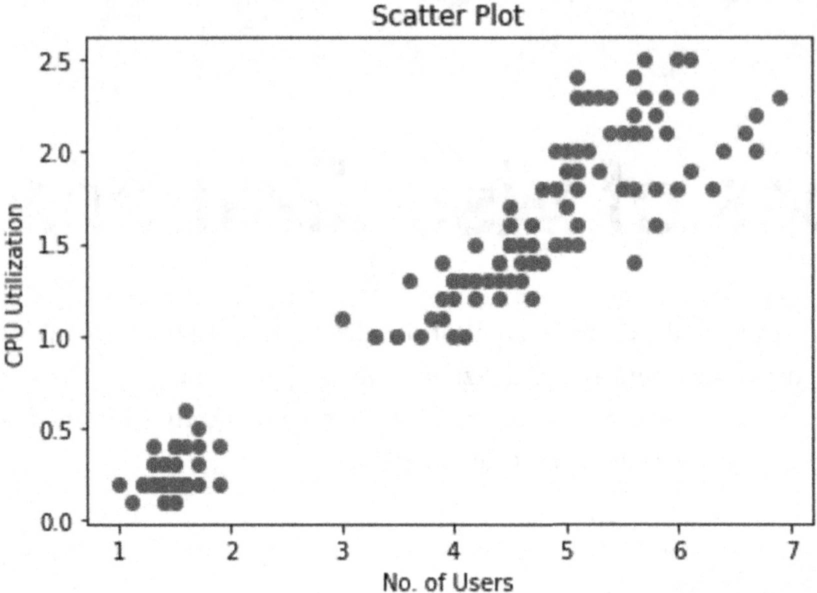


***Figure 16-1.*** *Scatter plot*

Figure 16-2 illustrates an example of a bar plot. This particular bar plot shows the CPU Utilization by Country. US has the highest CPU Utilization, followed by Canada and then Mexico.

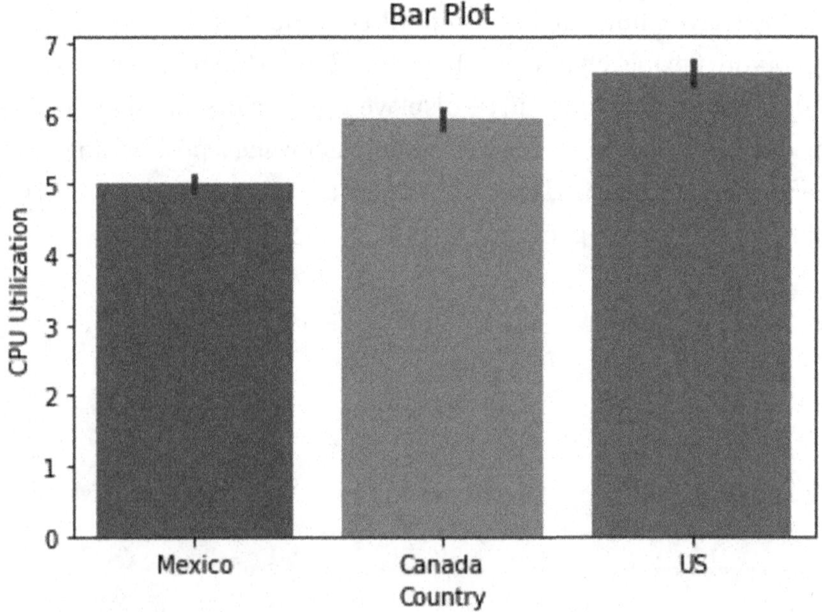


***Figure 16-2.*** *Bar plot*

A bar chart shows a categorical variable and the corresponding aggregate of another numerical variable.

Figure 16-3 illustrates a line plot. The line plot shows the CPU Utilization against the number of users. The CPU Utilization spikes when the number of users is between 5 and 5.5, and when the number of users is above 7.5.

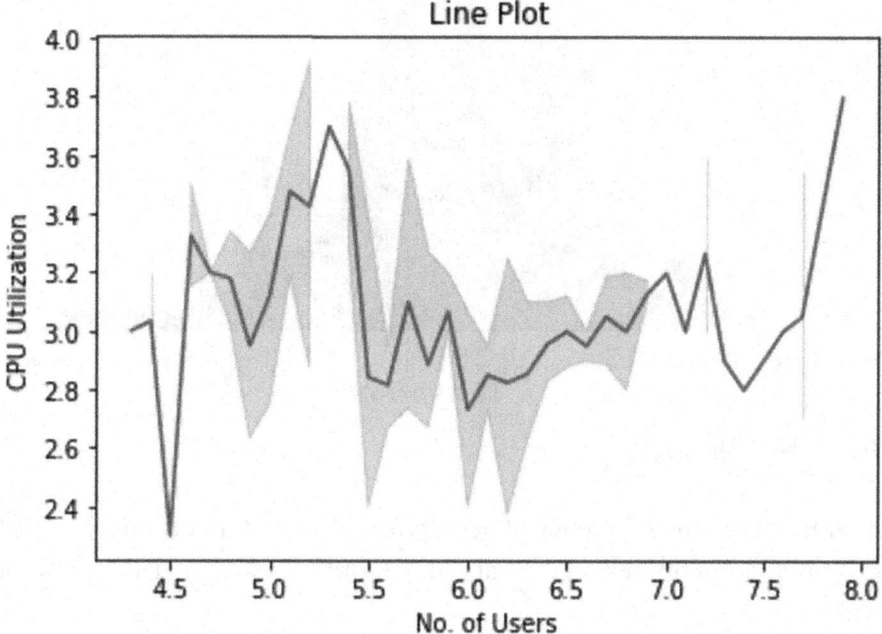


***Figure 16-3.*** *Line plot*

Line plots show the relationship between two continuous numerical variables. The shaded region displayed in Figure 16-3 shows the distribution of the y-variable given the value of x.

Figure 16-4 shows a histogram plot showing the frequency distribution of CPU Utilization. This plot shows that CPU Utilization has a normal distribution.

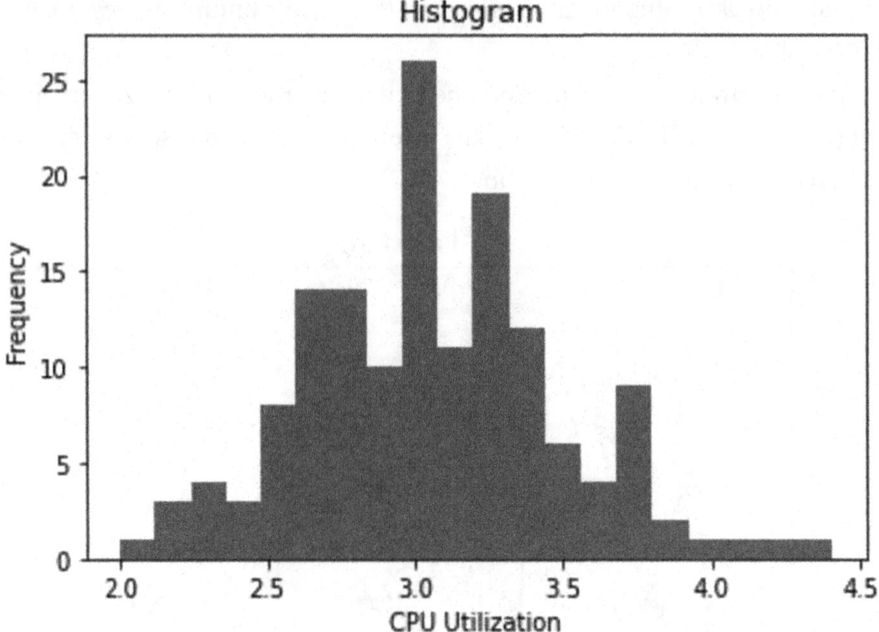


***Figure 16-4.*** *Histogram plot*

A histogram shows the frequency of occurrence of a categorical variable. The bin size determines the smoothness of the bars – the more bins specified, the smoother the distribution.

Figure 16-5 illustrates a pie chart showing the split of users' access types. For instance, most users have read access to the systems, followed by Read/Write Access. There is a much smaller number of users with Administrative privileges in the system.

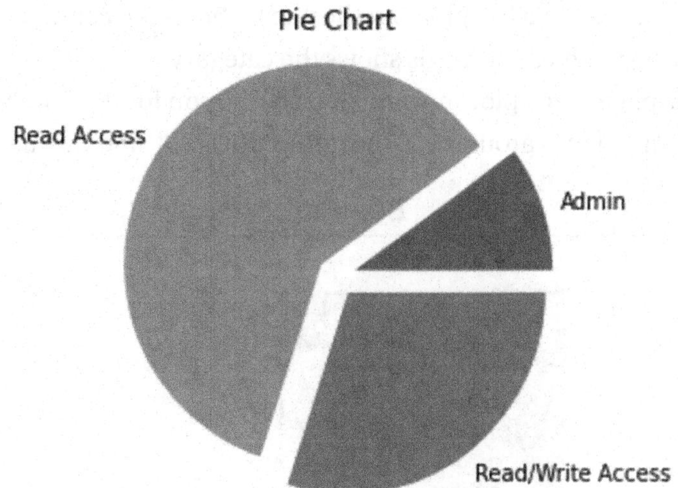


***Figure 16-5.*** *Pie chart*

A pie chart is used to display aggregated information for the categories within a categorical variable. It is better to use pie charts when there is a visible difference between the categories and a maximum of five categories.

Figure 16-6 shows a count plot of the Country field within the data set. The plot shows that there are 50 data points for each country, US, Canada, and Mexico.

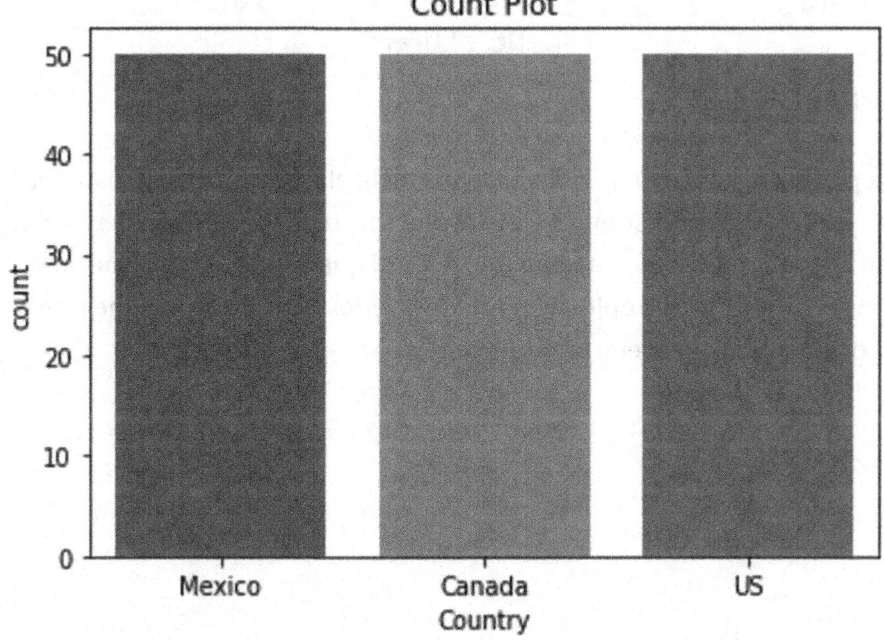


***Figure 16-6.*** *Count plot*

A count plot is similar to a bar plot, but the y-axis shows the count of the categories within a categorical variable. The x-axis shows the category.

Figure 16-7 displays a box plot showing the distribution for the number of users. In the given plot, the median is around 5.7, the minimum is 4.3, and the maximum is 7.9.

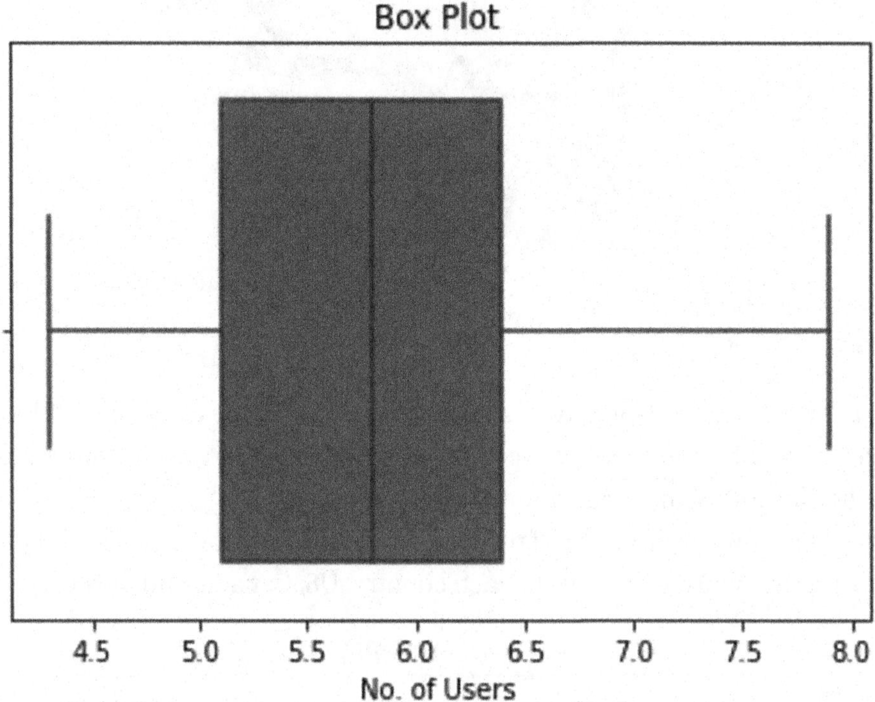


***Figure 16-7.*** *Box plot*

A box plot is one of the most valuable plots to display the distribution of a variable. It indicates the five-quartile summary in the plot (horizontal lines from left to right) – minimum, first quartile (25%), median (50%), third quartile (75%), and maximum.

Figure 16-8 shows the box plot with multiple series. This illustrates the true power of box plots of comparing between multiple series.

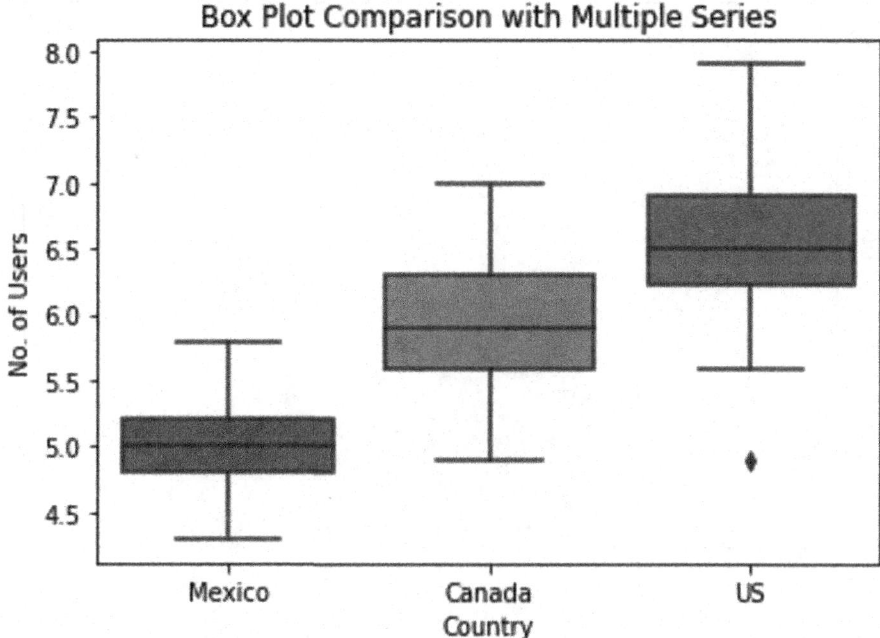


***Figure 16-8.*** *Box plot with multiple series*

The box plot can also be used to display the distributions of multiple categories within a categorical variable. Figure 16-8 shows that Mexico's median number of users is the lowest, followed by Canada and the US (the most median number of users).

Figure 16-9 shows a distribution plot for CPU Utilization. It is similar to the histogram plot. We can see that in the given plot, 3.0 has the highest CPU Utilization.

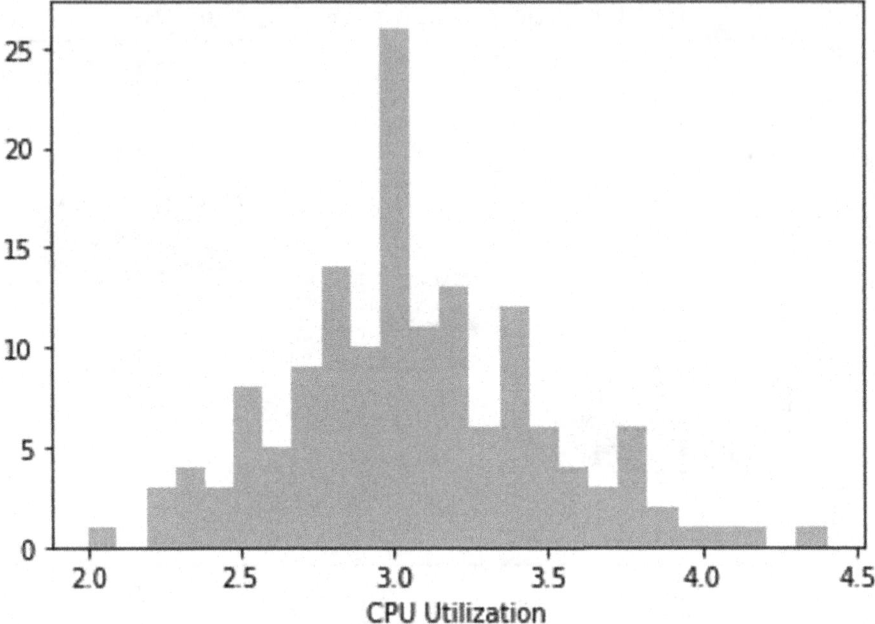


***Figure 16-9.** Distribution plot*

A distribution plot shows the distribution of a numerical variable. It can be used to confirm if a distribution is a normal distribution or skewed to the right or left. Groups of numerical data can be represented using a distribution plot for comparing range and distribution.

Figure 16-10 shows a distribution plot with a Kernel Density Estimate (KDE). KDE creates a smooth curve from the dataset. It can be used as a continuous representation of a discrete histogram plot.

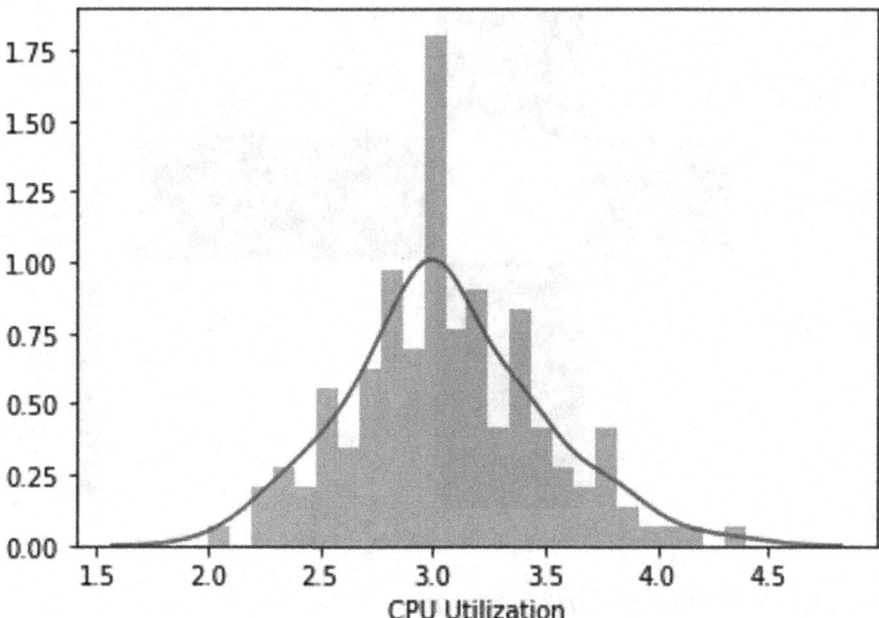


***Figure 16-10.*** *Distribution plot with kernel density estimate*

Figure 16-10 depicts the line representing the kernel density estimate superimposed on the bar plot. Creating this type of plot can be accomplished in Python using the 'kde=True' option in the distplot seaborn function.

# Advanced Visuals

Three advanced visualizations will be covered in this section. Heatmaps, Joint Plots, and Pair Plots are not used as much as the basic visualization discussed earlier. These plots can help visualize complex relationships between variables.

Figure 16-11 illustrates a heatmap showing the relationship between the variables in the data set. The correlations are represented in terms of colors that can interpret the level of correlation between the variables.

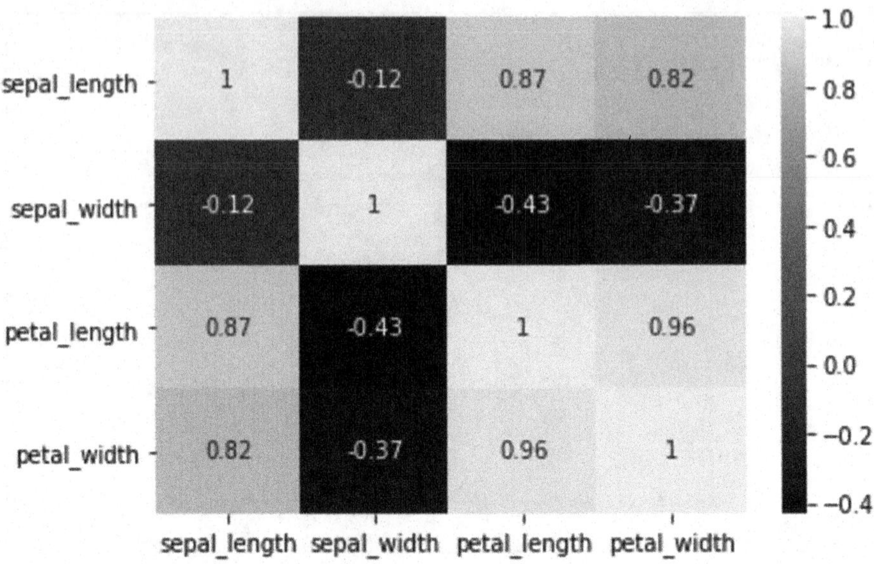


***Figure 16-11.*** *Correlation matrix heatmap*

A correlation matrix heatmap shows the correlation between two or more numerical variables. A categorical variable can be included in a heatmap, but it must be converted to a numerical variable using the one-hot encoding (or another encoding method). In Figure 16-11, petal_length has the highest positive correlation compared to petal_width (0.96). This means that as the petal_length increases, the value of petal_width goes up as well.

Figure 16-12 shows a joint plot showing the relationship between petal_length and sepal_length. A joint plot shows the scatter plot between two continuous variables and each of the variables (at the top and right, as shown in Figure 16-12).

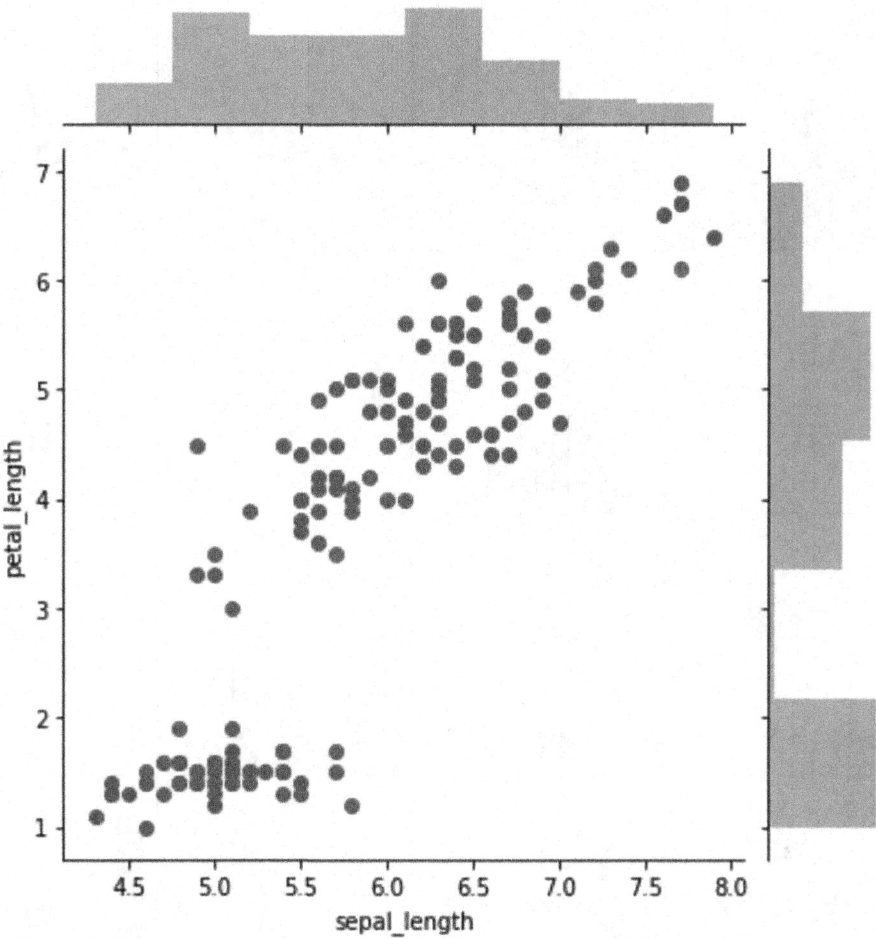


***Figure 16-12.*** *Joint plot*

In the plot, petal_length is distributed between 1 and 2, between 3 and 7. There is a break in the distribution between 2 and 3.

Figure 16-13 illustrates a pair plot between the variables in the data set. The given pair plot shows the relationship between four variables in the same plot.

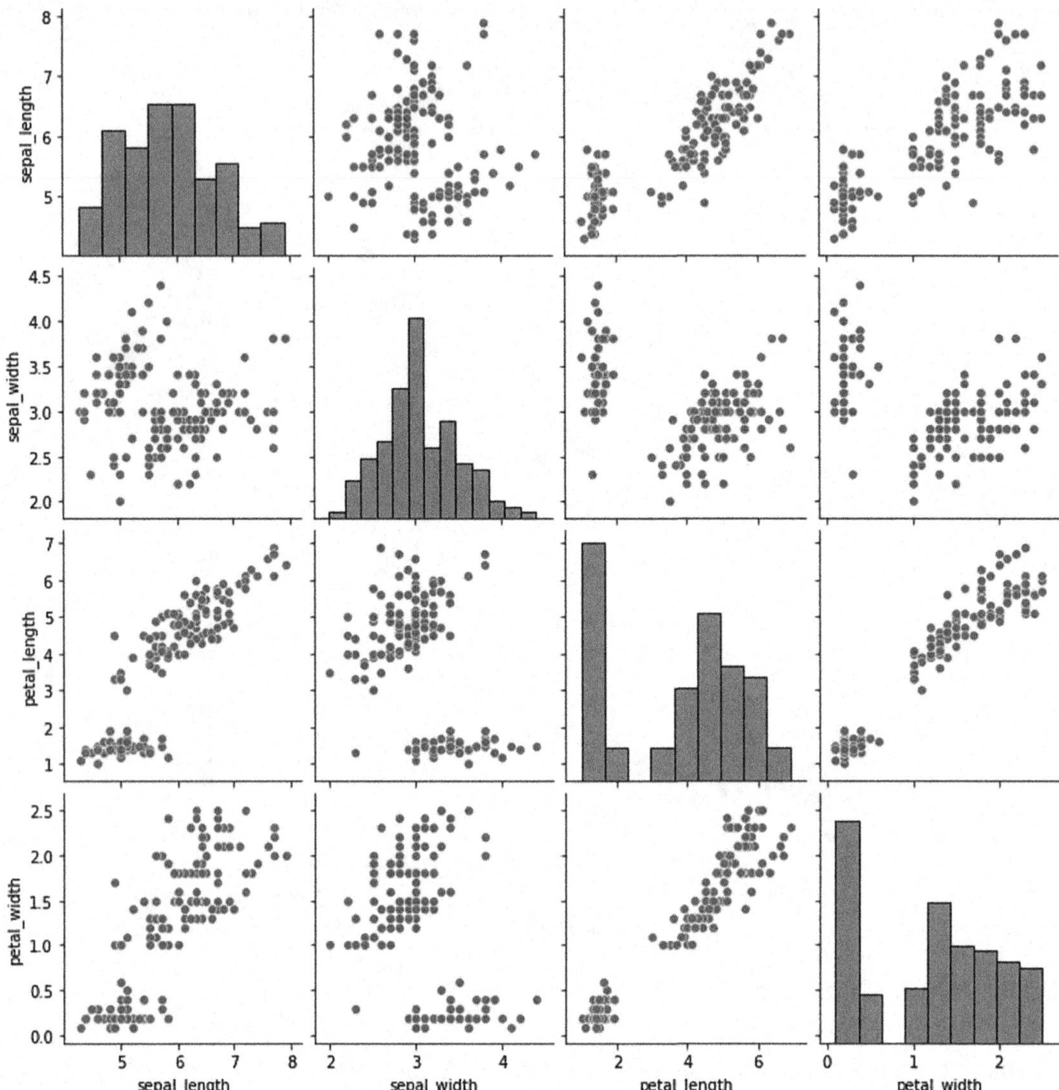


***Figure 16-13.*** *Pair plot*

The top row of Figure 16-13 shows how sepal_length is related to each of the other three variables, sepal_width, petal_length, and petal_width. The bottom row of Figure 16-13 displays the relationship between petal_width and the other variables, sepal_length, sepal_width, and petal_length.

## One-Hot Encoding

One-hot encoding can be used to convert categorical data into numerical data. This can be helpful when trying to find the correlation between a categorical variable and other numerical variables.

## Conclusion

The type of variable (numerical or categorical), the number of variables being compared, and whether the variables are discrete or continuous should point to the best visual to show the information. The features (variables) may need to be preprocessed before they are used in visualizations based on the message it is trying to convey.

# CHAPTER 17

# Effective Stories

A great deal of effort is required to create compelling stories. But a set of visualizations without a compelling story may not be enough to ensure that the reader grasps the key messages. In this chapter, we will explore three effective forms of storytelling using various mediums. The first visualization was given in a TED Talk, the second is a visual story, and the third visualization is presented on a web page. All the visualizations that are illustrated convey what we learned in previous chapters by incorporating elements that make a good data story. A recap of the essential elements that make a compelling story possible at the end of the chapter is listed for reference.

## Case Study: "The Best Stats You've Ever Seen"

Hans Rosling's TED Talk from 2006 is a classic when it comes to using data to captivate the audience. In this TED Talk, Rosling debunks myths or preconceived notions about the developing world using publicly available data from the United Nations. Rosling explains that he tested his college students on their views about the world by giving a pretest on Global Health. Rosling quizzed his students on questions such as "Which country has the highest child mortality?" and "Is there a gap between rich and poor?". After seeing the test results, he realized the students answered based on their preconceived notion that there is a division of the world into the Western World and the Third World. The child mortality rate and the number of people that are poor are much higher in the Third World than in the Western World. Rosling compiles publicly available data to verify the above assumptions based on actual data from the 1960s to early 2000s. He proceeds to show a scatter plot showing the relationship between Life expectancy and Fertility rate. World distribution of income, Child survival % vs. GDP per capita, and No. of Internet users vs. GDP per capita are other charts he shows to dispel myths on the divide between "Rich and poor" countries.



© Maris Sekar 2022
M. Sekar, *Machine Learning for Auditors*, https://doi.org/10.1007/978-1-4842-8051-5_17

Let us look (view the video link provided) at some of the tools Rosling uses to make his stories more effective, in addition to being a great presenter who is both humorous and charismatic. Rosling starts by introducing a conflict that also hooks the viewer into the presentation. His use of publicly available data adds to the credibility of his presentation. The graphs Rosling uses are engaging, although he mainly used two types of simple plots - scatter plots and distribution plots. He engages the audience by animating the scatter plots by the year of data available - this adds to the interaction and spontaneity of the presentation. The charts are all clearly labeled for the axes and titles. Only two dimensions of the data are presented simultaneously, except for shapes and colors used in the scatter plots to identify the countries and their population. The shapes and colors only made the charts more meaningful and did not complicate the charts. The key message was evident by the end of the presentation - there is no significant divide between the Western and Third Worlds when it comes to health and income.

Video Link: `www.ted.com/talks/hans_rosling_the_best_stats_you_ve_ever_seen`

## Case Study: "U.S. GUN KILLINGS IN 2018"

Periscopic is a socially conscious data visualization firm that posts insights to bring awareness to some everyday societal matters. One such visualization (view the video link provided) is its work on visualizing gun deaths in 2018. Once the web page is loaded, a projectile is seen shooting across the screen, and it appears to hit a virtual wall at a certain point, and then the projectile can be seen rapidly falling to what seems to be the ground. This projectile is followed by other similar projectiles that increase in number as the animation progresses. The path of the projectiles from their starting point to the time they hit a virtual wall is colored orange to signify the period of life of an individual. The path from the virtual wall to the end of the arc represents the life expectancy of that individual, and it is colored gray. This simple yet powerful visualization captures the number of years of life that was lost due to gun deaths in the United States in 2018, which, in this case, is calculated to be 472,332 years in total (shown in the top right corner of the screen). At the top left corner of the screen, the total number of people killed is shown to be 11,356 deaths. It also shows the distribution of the lifespan and the potential lifespan of the victims in two overlapping distribution plots.

The visualization is simple, with only two types of plots - the customized arc plot (explained above) and the distribution plots. The overlapping orange and gray colors show how the color of vibrant life fades into gray (conveying death). The critical message is hard to be missed – gun deaths take away many years of life. The animation of the projectiles shows the impact of multiple lives being lost without reaching the total expected lifespan. Visualizations can powerfully convey the effect and be impactful without being complicated.

Visualization Link: `https://guns.periscopic.com/`

## Case Study: "Numbers of Different Magnitudes"

Cole Nussbaumer Knaflic published a blog post in 2017 (see blog link) at *storytellingwithdata.com* showing how he transformed an initial graph that showed the risk profile of a loan portfolio. The original chart shows a bar graph that shows the breakdown of the risk profile over time. Beneath the bar graph, there are many numbers showing one year of data broken down by months. It is rather hard to derive any insight from the original chart. Cole then uses the same data but offers different visualizations to capture the insights in an easy-to-digest manner. He makes it easy to see that there was a significant increase in the portfolio when compared year by year. Cole derives at least 16 different insights from the same data using a series of visualizations. At the end of the analysis, he combines the most important insights into a one-page visualization that shows the status of the overall trend and the non-pass trend along with the corresponding visualizations.

Cole used only two types of visualizations – bar charts and line plots – to support all of his insights. Another striking feature of his plots is that they use action titles such as "81% increase in portfolio year over year," "Pass loans: from 89% to 95% of the total," and "Very high risk: decrease over time." Action titles help the reader to connect the dots so they can absorb the key underlying message more efficiently. The visualizations are color-coded based on the risk level. For example, "VERY LOW" risk level lines are gray, whereas "VERY HIGH" risk level lines are bright red. The colors make it easier to see how the different risk levels behave over time. Finally, the last one-pager visualization contains highlights that are important for the reader to see the most. "PASS increased from 89% to 95%" and "MODERATE RISK loans have more than tripled in volume in the past year." This is a great way to convey the message to a reader who does not have time to go through every visualization and insight to decode the key messages.

Blog link: `www.storytellingwithdata.com/blog/2017/4/19/numbers-of-different-magnitudes`

## Recap of Effective Storytelling Elements

Let us review the main elements of compelling storytelling, adding to what we learned in Chapter 14 and combine with what we learned in this chapter by looking at other effective visualizations.

**Here is a recap of the top recommendations for compelling storytelling:**

- Spend a significant amount of time understanding the business and the available data.
- Data quality needs to be verified before it can be used in visualizations.
- Remember to set the context for your data visualizations.
- Put the essential message/visual at the top left corner of the screen.
- Keep your visuals simple and clean (a maximum of two to three visualizations may be enough depending on the message).
- Use action titles to lead your audience to your key message.
- Use colors to distinguish categories.
- Use highlighted text to convey key messages.

## Conclusion

Visualizations are able to convey information at a glance. Stories, however, are the lens through which visualizations are interpreted. Combining the storytelling elements mentioned in this chapter with your visualizations helps you to make a strong impact on those viewing your results.

# CHAPTER 18

# Storytelling Tools

In this chapter, we will look at the technical effort required to make stories using visualization software. Some of the popular visualization tools will be discussed at a high level to give an overview of the visualization process. The general steps of the visualization process will also be briefly discussed.

## Technical Expertise

One of the misconceptions about producing storytelling visualizations is that you need a specialized skill set to make visualizations. Although a certain level of training is required, you do not require specialized training to get started with making compelling data stories. Most tools provide in-depth free training as part of the tools to help a beginner get started with their storytelling journey. As long as the key messages and requirements are known beforehand, one can build visualizations that support them.

One area where technical expertise is required is during the data preparation step. Data needed to develop the stories may not, in most cases, be in the format that can be easily consumed by the visualizations provided by the visualization platforms. For instance, suppose we want to make a bar chart showing the number of user logins for a system by day for the last year. Ideally, we want two fields, one for the x-axis or column and another for the y-axis or row. In this case, the column field is the date value for the last year, and the row field is the count of the number of logged-in users. The available data may not be in this format, to begin with, and may need to be processed to get the count of logins for each day.

Most auditors are familiar with MS Excel, and they perform their analysis in MS Excel as part of their testing process. In the next section, we will look at some other tools that are available to visualize data.



© Maris Sekar 2022
M. Sekar, *Machine Learning for Auditors*, https://doi.org/10.1007/978-1-4842-8051-5_18

# Available Tools

The visualization tools have very similar features. All of them support the visual types that are commonly used as standard elements. They provide an Integrated Development Environment (IDE) with drag-and-drop elements whose properties can be manipulated. Properties include color, size, position, title, format, and chart-specific properties such as legend position, axes, and data point properties. The visualization tools also provide the ability to combine multiple data sources, perform data preparation operations within the tool, and enable collaboration by sharing the visuals. Some of them have support for telling stories by providing annotation capabilities on top of the visuals. Let us look at some sample dashboards in a few visualization tools.

## Qlik

Figure 18-1 shows a visualization in the Qlik Cloud. There is a scatter plot on the left and a table showing the details of the customer sales on the right. The boxes on the top showing "Year," "Quarter," "Month," "Week," "Manager," and "Region" are filters for the visuals. For instance, you can filter the region for showing the data by using the "Region" filter.

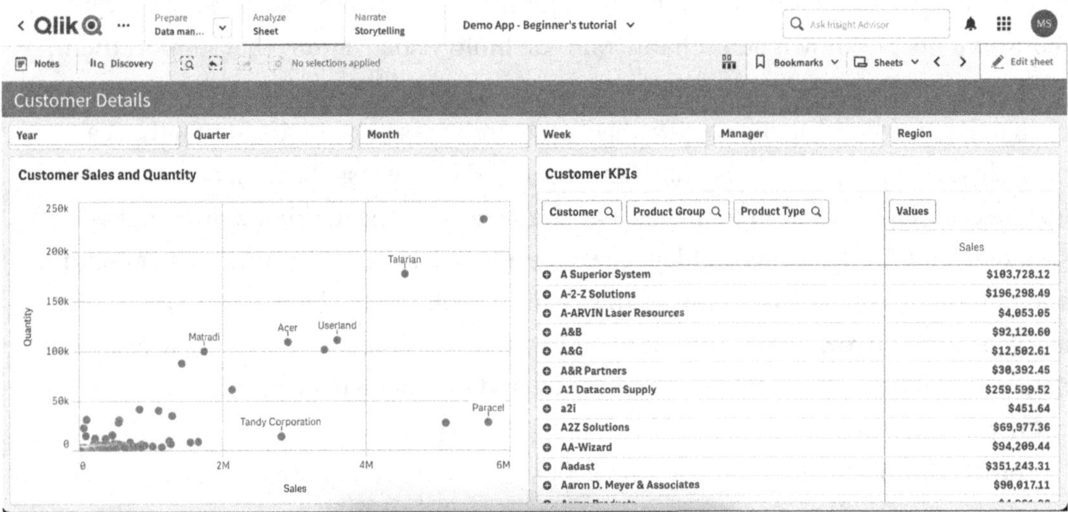


***Figure 18-1.*** *Qlik visualization*

# Power BI

Figure 18-2 shows a sample visualization in Power BI Desktop. It has a KPI chart, bar chart, ribbon chart, heat map, and a waterfall chart. The filters are available on the right tab named "Filters."

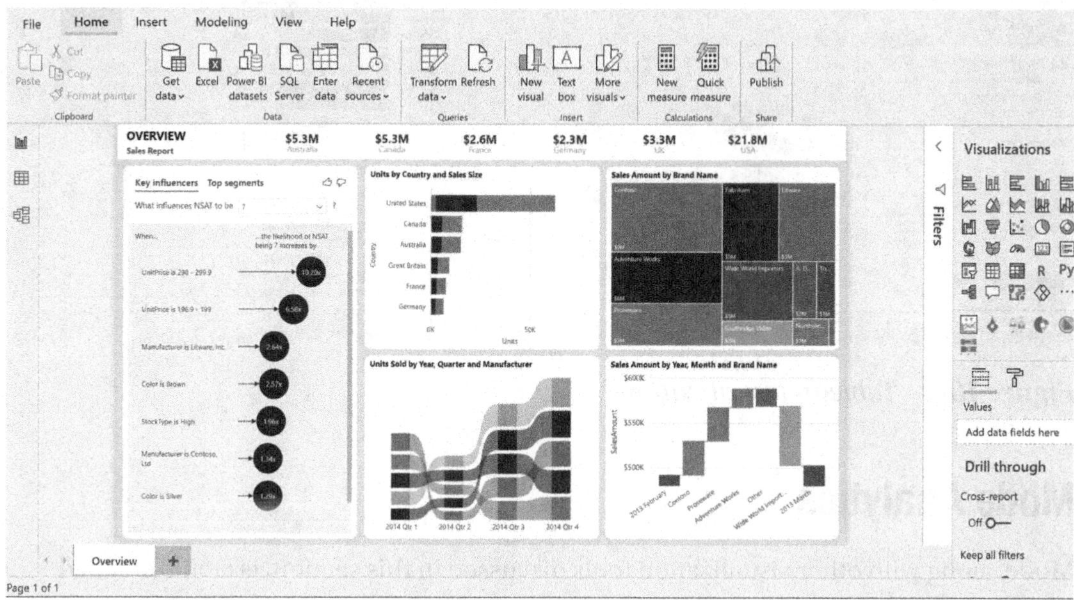


*Figure 18-2. Power BI visualization*

# Tableau

Tableau provides a similar interface to the others shown in Figure 18-3. Tableau, Qlik, and Power BI also have the ability to add customized visuals. Customized visuals help to extend the existing capabilities of standard visuals and create a new visual altogether. For example, you might want to make a custom visual that shows the organization chart. The classic visuals that come with the visualization software may not be able to support you with that. In this case, you might make a customized visual to display the organization chart and convey the message.

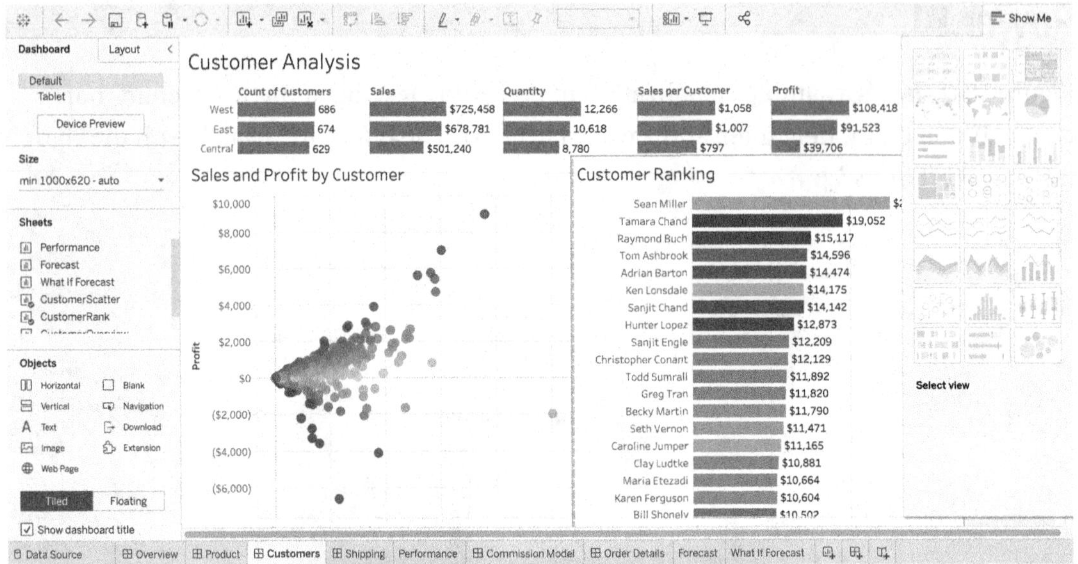


***Figure 18-3.** Tableau visualization*

# Mode Analytics

Mode, along with other visualization tools discussed in this section, is cloud-enabled. This means that we do not need to download a desktop version of Mode or Power BI to create visuals and share them with our colleagues. Users with a cloud account can log in to the visualization tool using a company portal account. Figure 18-4 depicts a visualization built with Mode.

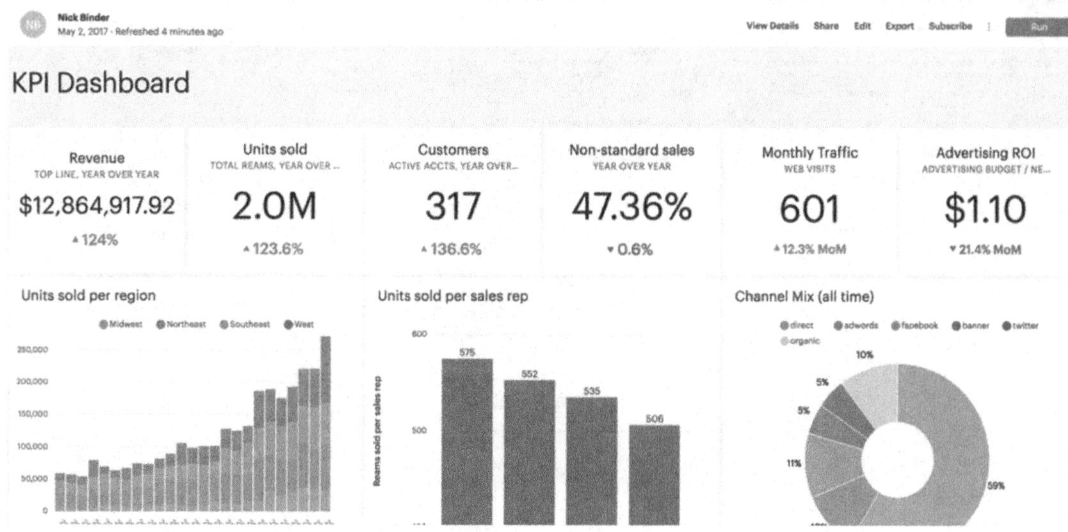


***Figure 18-4.** Mode visualization*

The tools discussed here are just a few of the popular ones.

Some visualization tools provide the ability to prepare and manipulate the loaded data within the tools themselves. For example, Power BI uses a built-in tool called the Power Query Editor that provides out-of-the-box data transformation/manipulation functionalities.

For instance, Figure 18-5 shows how a column type can be changed within Power BI using the Power Query Editor tool.

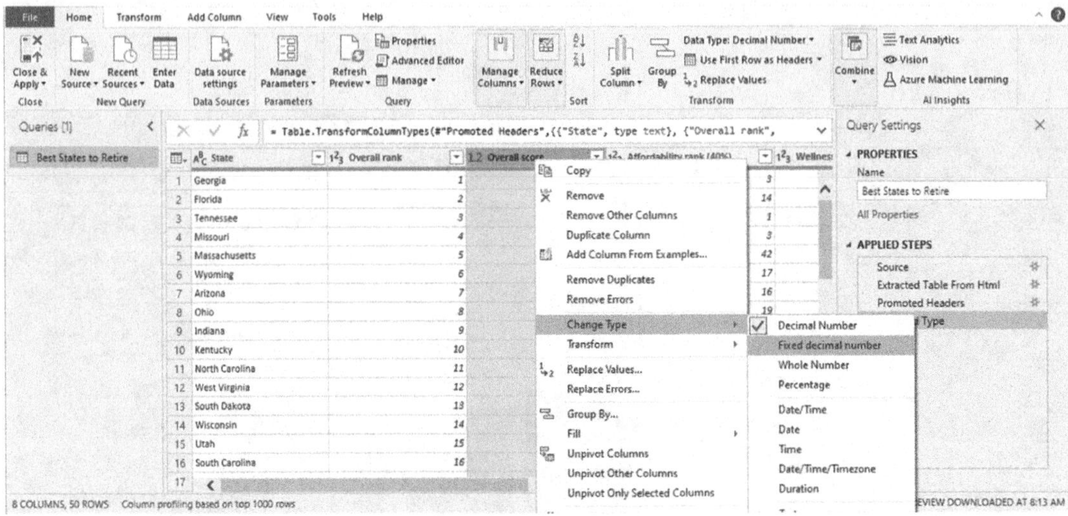


***Figure 18-5.*** *Data transformation using Power Query Editor*

Figure 18-6 shows a sample visualization environment setup with the steps discussed above.

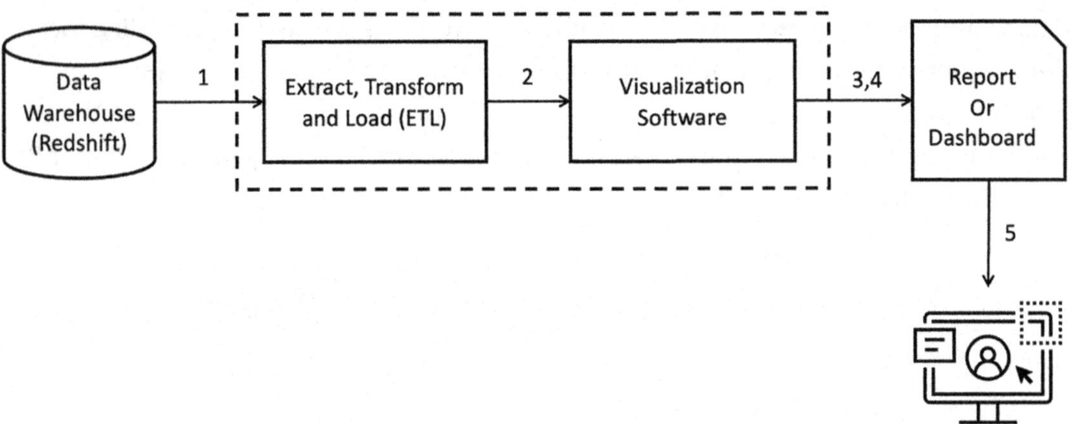


***Figure 18-6.*** *Visualization environment setup*

Although there are many visualization tools available, the general steps to creating visualizations powered by data are similar across them:

1) **Prepare data for visualization** – In this step, data is cleaned and formatted so that the data can be visualized.

2) **Extract/load prepared data into a visualization tool** - This step involves the loading of data into the visualization tool. Every tool provides the ability to load the data into the tool.

3) **Create visuals** - Visuals are created through drag-and-drop into the regions of choice based on the context and the message the story is trying to convey.

4) **Connect visuals with data** - After the visuals are placed on the worksheet, the properties of the visuals need to be modified. Properties include color of the data points, title, formatting, etc.

5) **Share visualization as a report or dashboard** - As a last step, the visualization with the story is shared with the end user. Some visualization software provides the ability to annotate on top of the visuals to help explain the story better.

# Conclusion

Most analytical tools provide an easy to use drag-and-drop interface that can be readily used to build visuals. Qlik, Power BI, Tableau, and Mode are some example tools that provide an Integrated Development Environment (IDE). Many analytic tools provide the ability to perform transformations within the tool as opposed to applying transformations before loading into the tool. The general steps to developing data visualizations are generally the same across all types of analytics tools.

# CHAPTER 19

# Storytelling in Auditing

In a previous chapter, the audit areas that are most suitable for storytelling in the audit process were briefly discussed. Some of the specific ways storytelling can be utilized in audits are explained in this chapter. Each of the use cases highlight the benefits of using storytelling.

## Audit Use Cases

Storytelling can be a powerful tool to confirm findings with the auditee and their business unit leaders. In addition, it can also serve as a way to explore other insights and to incorporate the domain knowledge learned over the course of the audit. There are three main types of use cases for applying storytelling in auditing:

1) Communicate findings in a simple and memorable fashion.
2) Support recommendations with credibility.
3) Confirm business knowledge.

# Communication of Findings

An auditor can use simple stories to communicate the findings based on the audit testing. Simple and relatable stories make it easy for auditors to connect with the audit report audience. The audit story, comparable to a creative story, can leverage context, the problem, action taken, and a result to build a strong narrative.

One of the effective stories to communicate a good audit story is to describe a problem in the context of the finding, how another organization or department dealt with the problem, and the outcome of the action. In order to address a similar finding in



© Maris Sekar 2022
M. Sekar, *Machine Learning for Auditors*, https://doi.org/10.1007/978-1-4842-8051-5_19

our organization, a comparable solution can be recommended in order to mitigate the risk. Stories are memorable due to their ability to connect with and move the emotions of the audience.

For instance, an organization struggles with keeping their systems updated on a regular basis. Even though the problem was highlighted in past audit reports for many years, no specific action was taken to mitigate the risk due to its perceived low impact on the systems. A story can be narrated showing how another organization that had a similar problem - not being able to keep their systems updated - was able to solve it. The organization had a cyberattack by a malicious actor (hacker). The hacker used a known vulnerability of an outdated operating system (Windows XP) to gain control over the organization's systems. By ensuring the systems are updated this attack could have been avoided altogether and the financial and reputational impact could have been reduced. This story is simple and effective in conveying the message and impact of the finding.

## Support Recommendations

Stories supported by facts and data analysis can give credibility to recommendations by explaining the "why" behind a recommendation. A recommendation is often included by the auditor along with the findings to talk about a suggested approach to mitigate the risk.

One of the ways data analysis and data science strengthens storytelling, further enhancing the recommendations, is by giving a holistic account of the state of a system. One hundred percent testing can be performed with exploratory data analysis techniques to give credibility to the story.

For example, when conducting an audit on reconciling vendor invoice amounts to the payment amounts, 100% of the invoices can be tested and highlighted in the story narrative. It may be seen that the amounts do not match for certain types of vendors. The story can highlight the fact that numbered vendor companies are more likely to charge above the amount specified on the invoice. This can add more credibility to the story and can help tailor an appropriate recommendation. The recommendation, in this case, can be to implement additional checks for numbered companies to ensure they are not overpaid.

## Clarify Business Knowledge

Audits can focus on highly specialized areas that internal auditors have not dealt with before. Auditors who are not familiar with the functional domain of the audit can confirm their understanding of business through their stories.

The stories can incorporate the business context or background to create additional impact. When the finding is described within the business context it may be more relatable to the auditee and the business unit.

Consider a case when a privileged access management IT Audit involves an access management software that the auditor has not seen before. The IT auditor assigned to the audit gains an understanding of the access management software through interviews with subject matter experts and other auditors that may be more familiar with the software. When sharing the findings of the audit, the IT auditor can craft stories that show how the findings can be reproduced within the software. For example, say the IT auditor finds that they are able to view personally identifiable information of all users in the access management software. It may be the case that due to elevated privileges of the IT auditor, they are able to see more information than a regular user can. This can be confirmed and/or clarified by the auditee when the IT auditor shares the story.

## Conclusion

Storytelling can be a powerful tool in various audit applications. The three main use cases of storytelling within Auditing are

- Communication of findings
- Explaining the “why” when stating recommendations
- Confirming business knowledge with domain expert.

# PART IV

# Implementation Recipes

# CHAPTER 20

# How to Use the Recipes

This chapter will contain details of how to use the recipes we present to get the expected result. It will outline the system requirements needed for the recipe, followed by where the reader can find the code. It will also talk about how the auditors can use the recipe and tweak it based on their needs. The implementers will need other granular details and may need to use the recipe in a different way than the auditors themselves, which will be discussed at the end of the chapter. Finally, implementation considerations are highlighted at the end of the chapter, including operationalizing the ML model.

## What Is a Recipe?

Recipes help the reader to apply what they have learned in this book in a cookbook format. Each recipe represents a type of problem an auditor is trying to solve in the real world. Some of the business problems faced by internal audit are explored in detail, and the corresponding solutions will be shared. Each recipe takes the format of what the end product will look like, followed by the required ingredients and how they will be used. The code is shared via GitHub. In order to ensure the implementation details are clear to the reader, the instructions are divided based on domain and technical expertise. If the auditor is not familiar with Python, they can provide implementation details to their internal IT department or an external consultant to help them implement it on their behalf. The solution will also be discussed from an audit point of view to ensure it is easy for the non-technical auditor to follow along.

Each recipe will consist of **four** sections:

- **The Dish.** This section defines the context and sets the background of the business problem. Benefits of implementing the proposed solution are also listed here.



© Maris Sekar 2022
M. Sekar, *Machine Learning for Auditors*, https://doi.org/10.1007/978-1-4842-8051-5_20

- **Ingredients.** This section will provide a listing of data inputs used and their detailed descriptions. The data sources and origin along with data preparation steps are listed here.
- **Instructions.** In this section, the procedures for using the ingredients are specified in a step-by-step manner.
- **Variation and Serving.** Some of the ways the solution can be modified to solve other problems is specified in this section. Lastly, the details about deploying the solution into a production environment is also described.

Although a thorough understanding of the models and their intuitions is not necessary for the auditor, they can refer to the "Serving" section for a general overview.

# Prerequisites

Python 3 and Jupyter Notebook are required to create and run the notebooks given in the recipes.

The recommended approach is to install both of them via Anaconda (`www.anaconda.com`). Anaconda installs Python, the Jupyter Notebook, and other commonly used packages for data science.

# Where Can You Find the Python Code?

Clone from this Git repository:

```
https://github.com/cube27/mlforauditors.git
```

You can use the following terminal command to clone the above repository:

```
git clone https://github.com/cube27/mlforauditors.git mlforauditors
```

## Implementation Considerations

When it comes to the implementation of the solution, there could be one of two scenarios:

1) The auditor is knowledgeable about the AI/ML technical domain.
2) Auditor leverages another team (internal or external) for their AI/ML technical expertise.

For the first scenario, the internal auditor is assumed to be familiar with the Python programming language, Jupyter Notebooks, and ML models in addition to their functional domain knowledge. An IT auditor with ML and Python background is an example of the skills expected for this case.

In the second scenario, consider the case when an auditor works with another team skilled in Python and/or ML. There could be multiple internal or external teams helping the auditor to implement the solution. An example of this would be an IT auditor working with the internal IT team for their Python expertise and leveraging an external consultant to help them with ML expertise.

In both scenarios, the auditor will incorporate their functional risk management and compliance know-how into the solution. Repeated infusion of domain expertise into the iterations of the models is essential to maximize the project's impact.

The CRISP-DM process is recommended to be used to implement the recipes. The solution, including the models and inputs, will need to be re-engineered based on the problem one is trying to solve. Even though solutions have been provided in the recipes, it is essential to focus on understanding the methodology rather than the code.

For instance, the next chapter contains the recipe for creating a Fraud and Anomaly detection ML solution. The code provided is the end product of iterations of going through the CRISP-DM process. It involves business understanding, the underlying data understanding, data preparation, modeling, evaluation, and finally, the deployment of the model into production. To leverage the exact solution for another organization, the related needs/problems of the organization need to be identified. The available data for the organization needs to be incorporated into the formation of the model. Additional data preparation steps may be involved based on the data quality. The provided model may need to be revisited to ensure it ties in with the new data. The performance of this new model will need to be evaluated and optimized. Through experimentation and various iterations, a final model will be produced.

The final model can then be deployed into the organization's production environment based on the resources of the internal audit team. A mature internal audit team with the ability to collect and analyze data may be able to put the model into production themselves. An internal audit team that does not have these resources may need to engage IT Operations to help them operationalize the solution. Once the model is in production, internal audit will need to develop a process to identify and follow-up on exceptions generated using the Fraud and Anomaly Detection model. The model itself will need to be continuously monitored for performance to identify and correct model drift through optimization. Model drift is the degradation of model performance over time due to changes in data and relationships between the input and target variables.

Figure 20-1 shows the CRISP-DM process for data science that was introduced in an earlier chapter.

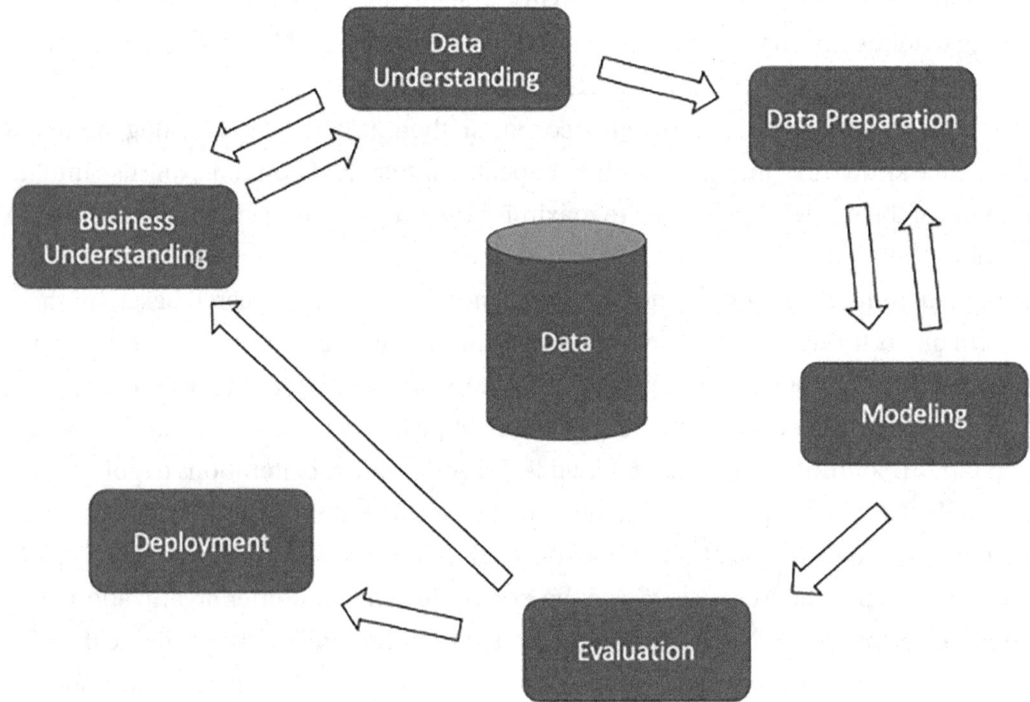


***Figure 20-1.*** *CRISP-DM framework for data science projects*

A Proof of Concept (POC) is first recommended to be built using the provided recipes before operationalizing and moving the final models into production. The POC can use data with limited scope to iterate and get to a minimum viable product (MVP) as quickly as possible. A POC is necessary before spending time and resources on the solution so that the benefits and effort needed to build the solution are clear earlier on. A stage-gate review process can be utilized to implement a more detailed solution with all the available data if the benefits outweigh the costs.

# Conclusion

The use cases provided in Chapters 21 to 27 show the utilization of ML in common audit applications. The problems and their solutions are similar to cooking recipes. They are organized into the following sections: the dish, ingredients, instructions, variation, and serving. The solutions are accompanied by Jupyter Notebooks that are available via a publicly available GitHub repository. It is recommended that a Proof of Concept (PoC) with a smaller data set and a simple model be first built before putting the models into production. This is important because not all models may be eligible for running in production environments. The benefits of running the model in production must outweigh the resources and costs needed to put the model into production.

# CHAPTER 21

# Fraud and Anomaly Detection

In this chapter, an intelligent fraud and anomaly detection system using statistical, supervised, and unsupervised machine learning techniques is presented. The following sections will show how a non-traditional approach can be leveraged by internal auditors to gain additional insight into the data. The chapter is organized like a recipe – goal of the recipe, ingredients, instructions, and variation and serving. The accompanying code is available at the GitHub repository specified in Chapter 20.

## The Dish: A Fraud and Anomaly Detection System

The dictionary defines Fraud as "the intentional perversion of truth in order to induce another part with something of value." Fraud is becoming increasingly complex and harder to detect in organizations. As mentioned in earlier chapters, COVID has increased the likelihood of fraud due to new avenues of exploitation, including remote work. There are multiple news articles of fraud investigations being launched in many regions of the world.

In organizations, there are many areas where fraud can potentially occur, including HR, Finance, and Supply Chain Management. In this recipe, we will focus on payment fraud related to payments made to the vendor. When an organization pays its vendor for the work received, there are checks put in place to prevent fraud. Most of these checks are manual controls such as verifying the identity of the vendor when making the payment and ensuring that a larger than usual amount is not being forwarded to the vendor. There are automated controls provided by Enterprise Resource Planning (ERP) systems included in software such as SAP and Oracle. The automated controls range from restricting field values at the point of data entry to more complex checks such as the three-way match reconciliation between purchase order, goods receipt note, and



© Maris Sekar 2022
M. Sekar, *Machine Learning for Auditors*, https://doi.org/10.1007/978-1-4842-8051-5_21

invoice. Unfortunately, fraud still occurs with all these controls being present. Internal auditors conduct audits to detect fraud but are often restricted to testing based on sampling or simple filtering techniques. For example, certain keywords can be searched and filtered in the payment transactions to determine if a payment is fraudulent.

Anomaly detection systems provide a way to minimize the risk of fraud further by looking at anomalous behavior in the data. In machine learning, anomaly detection systems can employ both supervised and unsupervised learning techniques to detect red flags in the data. As noted in a previous chapter, the mere detection of red flags does not constitute a fraud. Each red flag needs to be investigated on a case-by-case basis and confirmed with the business (possibly the Finance department).

The goal of the anomaly detection system is to find the outliers observed in the dataset using an algorithm. An interesting aspect of this is most anomaly detection systems work with imbalanced datasets. There is a far smaller number of outliers compared to the majority class. For example, in real-life scenarios for fraud detection, only a handful of transactions are thought to be fraudulent. This poses some additional challenges that can be overcome through the selection of models.

In this recipe, we will prepare three ways of detecting anomalous behavior in SAP vendor payment transactions:

1. Interquartile method (statistical)
2. Support Vector Machines - Supervised machine learning
3. k-means - Unsupervised machine learning

The cells can be run one by one in the Jupiter Notebook by clicking the "Run" button in the top toolbar. The cells can all be run at the same time from top to bottom by going to the "Cell" menu at the top and clicking "Run All."

# Ingredients

In addition to the prerequisites defined in Chapter 20, you will need the following data for this recipe:

- BKPF SAP Table - Header data for all financial transactions
- BSEG SAP Table - Segment data for all financial transactions

We will need a combined table that joins BKPF and BSEG. The first five transactions in the table are shown in the following figure.

| BUKRS | GJAHR | MONAT | USNAM | WAERS | BSCHL | KOART | DMBTR | HKONT | LIFNR |
|---|---|---|---|---|---|---|---|---|---|
| 1000 | 2018 | 1 | MSEK | CAD | 25 | K | 200 | 100000 | 11223344 |
| 1000 | 2018 | 2 | MSEK | CAD | 25 | K | 200 | 100000 | 11223344 |
| 1000 | 2018 | 3 | MSEK | CAD | 25 | K | 200 | 100000 | 11223344 |
| 1000 | 2018 | 4 | MSEK | CAD | 25 | K | 200 | 100000 | 11223344 |
| 1000 | 2018 | 5 | MSEK | CAD | 25 | K | 200 | 100000 | 11223344 |

Table 21-1 displays the listing of the columns along with their descriptions.

***Table 21-1.*** *Column listing of FI_Transactions.csv*

| Column | Description | Table |
|---|---|---|
| BUKRS | Company Code – The entity within the organization the transaction is part of | BKPF |
| BELNR | Document Number – Unique identifier of the document | BKPF & BSEG |
| GJAHR | Fiscal Year – The year of the transaction | BKPF |
| MONAT | Fiscal Month – The month of the transaction | BKPF |
| USNAM | Username – The user who entered the transaction | BKPF |
| WAERS | Currency Code – The currency of the transaction | BKPF |
| BSCHL | Posting Key – This field is used to distinguish between the various postings in SAP – invoice, payment, credit, etc. | BSEG |
| KOART | Account Type – There are many account types based on if the general ledger or subledger is used | BSEG |
| DMBTR | Amount in Local Currency – This is the amount of the transaction in the local currency | BSEG |
| HKONT | Account Number | BSEG |
| LIFNR | Vendor Number | BSEG |
| Fraudulent | Created by us to label fraudulent transactions we know from past history. 1 represents a non-fraudulent transaction and -1 means this transaction was confirmed to be fraudulent. | N/A |

Note that while extracting BSEG and BKPF from SAP, we do not need all the columns for this recipe. We just need the columns from the preceding corresponding table. The "Fraudulent" column is added by us to the combined table.

# Instructions

## Step 1: Data Preparation

The only data preparation in this analysis is importing the joined BSEG/BKPF into the Notebook. The tables can be joined outside of the Notebook, or it can be done within Python.

To do it within Python, use the following command, given that BSEG and BKPF have been loaded into the notebook via the `read_csv()` function:

```
BKPF.join(BSEG.set_index('BELNR'), on='BELNR')
```

The BELNR field is common to both BSEG and BKPF.

There are some minor data transformations needed before the models are trained. This will be highlighted in the Notebook.

## Step 2: Exploratory Data Analysis

This section helps to explore the data using some of the readily available tools within Python. The structure and the actual data distribution are analyzed to give the reader a deeper understanding of the provided data. Two of the most useful functions are `describe()` and `unique()`.

```
# Summary statistics for all columns
df.describe(include='all')
```

| | BUKRS | BELNR | GJAHR | MONAT | USNAM | WAERS | BSCHL | KOART | DMBTR | HKONT | LIFNR | Fraudulent |
|---|---|---|---|---|---|---|---|---|---|---|---|---|
| **count** | 50.000000 | 5.000000e+01 | 50.000000 | 50.000000 | 50 | 50 | 50.0 | 50 | 50.000000 | 50.000000 | 5.000000e+01 | 50.000000 |
| **unique** | NaN | NaN | NaN | NaN | 4 | 2 | NaN | 1 | NaN | NaN | NaN | NaN |
| **top** | NaN | NaN | NaN | NaN | MSEK | CAD | NaN | K | NaN | NaN | NaN | NaN |
| **freq** | NaN | NaN | NaN | NaN | 21 | 40 | NaN | 50 | NaN | NaN | NaN | NaN |
| **mean** | 1340.000000 | 2.000003e+07 | 2019.640000 | 6.340000 | NaN | NaN | 25.0 | NaN | 6344.000000 | 126000.000000 | 1.122335e+07 | 0.800000 |
| **std** | 688.387646 | 1.457738e+01 | 1.102132 | 3.230025 | NaN | NaN | 0.0 | NaN | 23904.189369 | 44308.749769 | 6.456258e+00 | 0.606092 |
| **min** | 1000.000000 | 2.000000e+07 | 2018.000000 | 1.000000 | NaN | NaN | 25.0 | NaN | 100.000000 | 100000.000000 | 1.122334e+07 | -1.000000 |
| **25%** | 1000.000000 | 2.000001e+07 | 2019.000000 | 4.000000 | NaN | NaN | 25.0 | NaN | 200.000000 | 100000.000000 | 1.122334e+07 | 1.000000 |
| **50%** | 1000.000000 | 2.000003e+07 | 2020.000000 | 6.500000 | NaN | NaN | 25.0 | NaN | 200.000000 | 100000.000000 | 1.122334e+07 | 1.000000 |
| **75%** | 2000.000000 | 2.000004e+07 | 2021.000000 | 9.000000 | NaN | NaN | 25.0 | NaN | 800.000000 | 175000.000000 | 1.122335e+07 | 1.000000 |
| **max** | 5000.000000 | 2.000005e+07 | 2021.000000 | 12.000000 | NaN | NaN | 25.0 | NaN | 100000.000000 | 200000.000000 | 1.122337e+07 | 1.000000 |

***Figure 21-1.** Summary statistics shown by describe( ) function*

The describe function provides some summary statistics such as mean, mode, quartiles, including the min, max, and standard deviation, as shown in Figure 21-1. For the given dataset, it is illustrated that the amount has a min, 25%, 50%, 75% and max as 100, 200, 200, 800 and 100000, respectively. This suggests that there is a right skew to the distribution since the 75% quartile is about four times higher than the previous two quartiles. Also, there appears to be an outlier based on the max value of 100000, which is much greater than the mean and median.

The unique function gives a helpful list of all possible unique values of the categorical field observed in the dataset. In the given dataset, there are four total users (MSEK, KMES, PKIM, LUKE) and two currencies (USD, CAD).

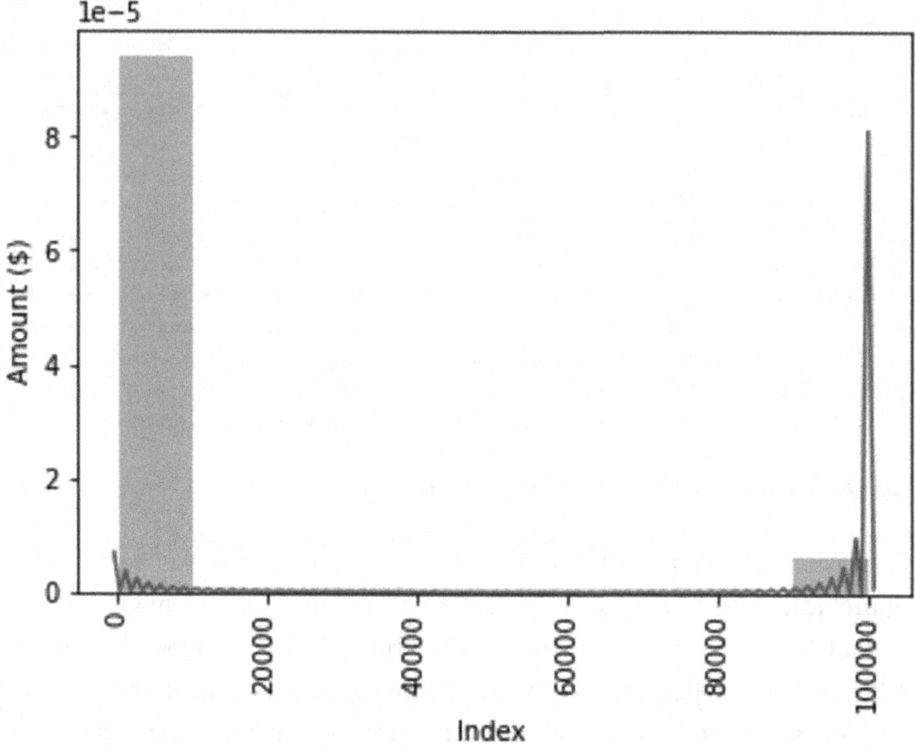


***Figure 21-2.*** *Histogram showing the amount distribution*

Finally, the amounts are plotted in a histogram as shown in Figure 21-2 to see how the amounts are distributed in the dataset. The histogram shows that most of the amounts are between 0 and $ 10,000 and there are very few data points around $ 100,000.

## Step 3: Apply Interquartile Range (IQR) Method

The IQR method uses a statistical approach to detect outliers. The interquartile is defined as the range between the first quartile (25%) and the third quartile (75%). Most of the data is set to be distributed in this window for a normal distribution. The given sample dataset is not a normal distribution, as can be observed from the histogram. The method works best for normal data distributions; it is shown here for demonstration so that the reader can leverage it for their own dataset.

The suspected fraudulent transactions are compared with the ones known to be actual fraud transactions. At the end of the analysis, the classification report shows that there is 96% accuracy (F1-score) when it comes to the algorithm's effectiveness in detecting non-fraud transactions. For fraud transactions, the accuracy goes much lower, to 50%. This means that it is much harder to detect an anomaly than it is to say all transactions are not fraudulent transactions. This makes sense due to the imbalanced nature of the dataset. If the dataset is imbalanced, SMOTE can be used to balance the dataset out. SMOTE stands for Synthetic Minority Over-sampling Technique and is an oversampling technique that creates synthetic minority class samples to make up for the imbalance.

## Step 4: Perform Supervised Learning

The next fraud detection algorithm uses supervised machine learning, specifically, a one-class Support Vector Machine (SVM) model to detect anomalies.

The target and features are first defined. This is followed by the splitting of the dataset into training and test sets, as is characteristic of all supervised machine learning algorithms. A 50/50 train/test split is specified due to the small number of transactions. There are only 50 transactions in the given dataset. For a dataset with >10,000 transactions, a 90/10 train/test split is recommended, that is, 90% of the dataset is chosen for the training set and the remaining 10% data is used for the test set. This is important for the model to generalize well on the dataset.

The SVM model has a slightly lower accuracy of 95% for detecting non-fraudulent transactions and a much higher accuracy of 67% when it comes to correctly identifying fraud. Hence, the SVM method appears to be doing much better compared to the previous method.

## Step 5: Perform Unsupervised Learning Analysis

For the next analysis, we use the unsupervised learning algorithm K-Means Clustering. Normally, K-Means Clustering is used to identify hidden insights in the data without specifically asking the algorithm for a question. Clusters or themes within the data are identified and analyzed to see if there is a pattern within the features.

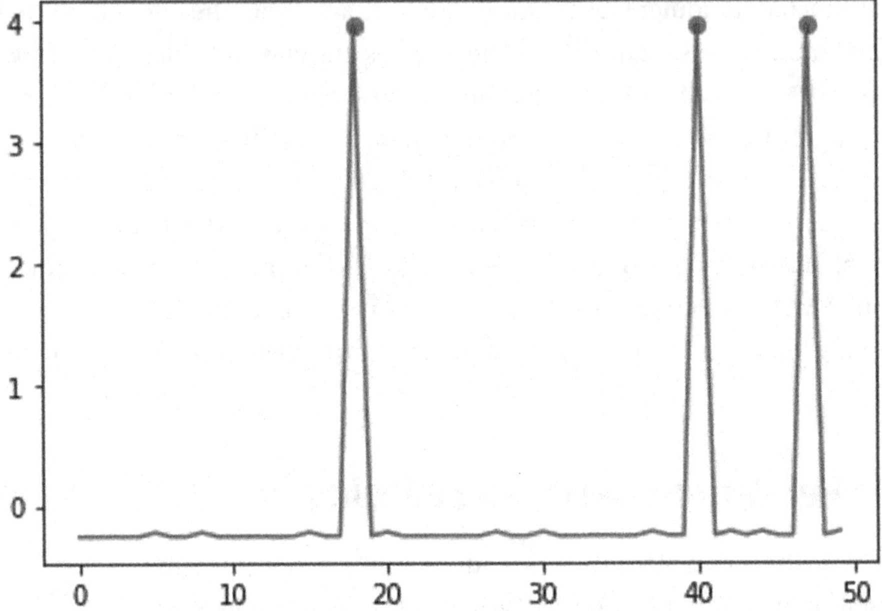


*Figure 21-3. Anomalies after performing K-Means Clustering*

In this implementation, k-means is used to identify a single cluster that is at the center of the dataset provided. The Euclidean distance is then used to determine the distance between the center of the cluster and all the remaining data points. The furthest three points are then identified as the anomalies in the dataset. The anomalies have been highlighted in red, as shown in Figure 21-3.

The data is first scaled because K-Means performs better with normalized data. The anomalies are identified as red points on the final plot. After this, the labels for the predicted anomalies are set as -1 and the evaluation metrics are shown. The ability of the algorithm to detect non-fraudulent transactions is again higher (96%). The f1-score for the algorithm to detect fraudulent transactions is 50%. These results are similar to the IQR Method, but are less accurate when compared to the second SVM method.

## Step 6: Review Exceptions with Additional Data

In this analysis, the actual fraudulent transactions were identified before performing the analyses. The fraudulent transactions were provided to evaluate the performance of the ML models using the data. After the models have been evaluated, the final models can be used on unknown transactions to detect fraud. The exceptions will then be forwarded to the proper business unit (possibly the Finance team in this case) for confirmation of fraudulent activity.

## Step 7: Re-evaluate the Models

The final models will need to be re-evaluated on a regular basis to check for their performance with the provided data. This can be done by getting the evaluation metrics, particularly, the f1-scores of the models. If there is a deterioration in the performance, the parameters for the models will need to be fine-tuned or the data will need to be further processed.

# Variation and Serving

The same code can be leveraged for other ERPs through relinking the fields. For example, in SAP, the amount value is stored in the DMBTR field in the BSEG table. The amount in local currency in another ERP can be determined and remapped. Some additional data preparation steps may be needed to ensure that the correct currency is being used. Some ERPs provide amounts in foreign currencies only. In this scenario, the amounts will need to be converted to a base currency (the local currency of the company) before it can be fed into the models.

There is a wide variety of features available in the ERP. In SAP, the BSEG and BKPF tables have over 100 fields each. Not all fields may be relevant for the analysis being used. There are many ways features can be evaluated for feature selection. One such method is Principal Component Analysis (PCA), which was discussed in an earlier chapter. PCA reduces the number of features and helps us avoid redundant features that do not add additional value.

One way to leverage dates in the current analyses is by calculating the difference in dates to get another feature that can be used in the analysis. For example, the K-Means algorithm requires all features to be numerical. In order to utilize the information contained in the dates, a new field that shows the difference between the payment date and the invoice date can be calculated and used as a feature for the K-Means algorithm.

For the SVM and k-means models, the Jupyter Notebook code only uses the amount (DMBTR) and the vendor number (LIFNR) fields. Other fields may be used for the features. In both of the cases, it was found that additional features negatively affected the performance of the models based on the provided sample data. This may not be the case with real data. It is important to change the features based on the data being used and re-evaluate the model to check for its effect on the model's accuracy. Without going into details, SVM is said to perform better with high-dimensional data, and K-Means Clustering suffers from the curse of dimensionality. This means that the performance of k-means is worse for a larger number of features when compared to the SVM model.

For SVM, the algorithm uses a 50/50 split for train/test data. As mentioned earlier, a 90/10 split for train/test dataset is suggested for datasets greater than 10,000 rows.

Ideally, the final model with the greatest performance will be deployed into production and consumed by other areas of the business. Many cloud providers such as Amazon Web Services (AWS) and Google Cloud Platform (GCP) provide ways to deploy the final models and expose them as an endpoint that can be consumed by other applications in the business. This method helps reduce costs related to supporting the infrastructure needed to deploy the models.

Depending on the frequency of the fraud checks, the models can be evaluated on a quarterly, monthly, or weekly basis based on the industry. For banking applications, a more real-time or continuously streaming data application may be needed. For audit applications, a quarterly run of the model may be sufficient. As stated earlier, it is essential that the models are continuously evaluated for their performance since the data being used can change over time which can further cause the model to drift. This results in low performance that works against the ROI of the models.

# CHAPTER 22

# Access Management

This chapter shows the application of a scatter plot to support an access management audit. The security data from SAP is used to show how the access of all people in the organization can be evaluated using a novel approach. The provided use case can be easily leveraged by auditors for other ERPs due to the generalized nature of the solution. The chapter is organized like a recipe – goal of the recipe, ingredients, instructions, and variation and serving. The accompanying code is available at the GitHub repository specified in Chapter 20.

## The Dish: ERP Access Management Audit

Access management audits are regularly performed by internal auditors to ensure the organization is comfortable with the way it assigns, revokes, and maintains access for its people. This is a key control in ensuring only authorized personnel have access to the information they need. The general policy is to provide access only on a "Need to Know" basis. Only the HR team can see salary information for others in the organization. Historically, the access management audits are performed using spreadsheets. Due to the size of the organization, auditors often resort to sample-based testing. In this case, administrators with elevated privileges and a sample from each of the roles in the organization are tested to verify if they belong to the intended access group within the ERP.

There are two main types of access controls: Role Based Access control (RBAC) and User Based Access control (UBAC). In RBAC, which is the preferred type of access control, the system role is assigned based on the functional area of the employee. The supply chain team has access to supply chain info, but auditors need read only elevated privileges to perform their audits are example of RBAC. In UBAC, the individual user is assigned an access if they need the info, but do not specifically belong to that business unit. Ideally, RBAC is widely used for almost all employees and UBAC may be used to



© Maris Sekar 2022
M. Sekar, *Machine Learning for Auditors*, https://doi.org/10.1007/978-1-4842-8051-5_22

assign individuals in the organization with special privileges. For instance, a supply chain team member may need to retrieve anonymized HR info to calculate the cost of labor. In this case, a special access may be assigned to the supply chain team member to do their job. In this audit, we focus on RBAC, but it can also be used to identify or confirm gaps in UBAC for the organization.

# Ingredients

You only need access to the raw table named "AGR_USERS" in SAP to conduct this analysis. This table contains the user role assignment for users within SAP. For example, the first record in the AGR_USERS table illustrates that user "MSEK" has "GENERAL" access to SAP. Table 22-1 shows additional details for all the columns.

***Table 22-1.*** *Column listing of AGR_USERS.csv*

| Column | Description |
|---|---|
| AGR_NAME | The role assigned to the user |
| UNAME | The username of the SAP user |
| FROM_DAT | The date the user was granted access to the role |
| TO_DAT | The date the access was revoked from the user. |

# Instructions

## Step 1: Data Preparation

The only major transformation is the addition of a new column called "Access Days." The GitHub code includes this transformation. The transformation is done through a function called `date_convert()`. Date_convert takes in a date in string format and converts to the datetime format in Python. The date is then subtracted from the date the analysis was performed. The difference provides the number of days a user had access to the corresponding role. For example, on line two in the AGR_USERS table, it can be seen that user "MSEK" had access to "FINANCE" role from January 15, 2017, and this privilege has not since been revoked – this means that he had access to this role for 1764 days (based on the run date of Nov 14, 2021).

## Step 2: Exploratory Data Analysis

The size of the data is retrieved first as part EDA. The sample data provided contains a total of 25 rows and four columns. The summary statistics shows that "GENERAL" access role is assigned to all the five users. This appears to be a general access provided to everyone with access to SAP. January 15, 2017, had the most number of role start dates. All the roles have an end date of "9999-12-31," which is a general convention in SAP to indicate that it has not been revoked yet. If there is a date provided other than "9999-12-31," it means the role access ended on that date. There are seven unique roles defined in the system – one for each function in the organization like "FINANCE," "IT SUPPORT," "HR SUPPORT," etc.

## Step 3: Scatter Plot of ERP Access

In the final scatter plot of the analysis, the roles are indicated in the y-axis and users are placed in the x-axis. The size of the bubbles indicates the number of days the user had access to the corresponding role. The color of the bubbles indicates the business function.

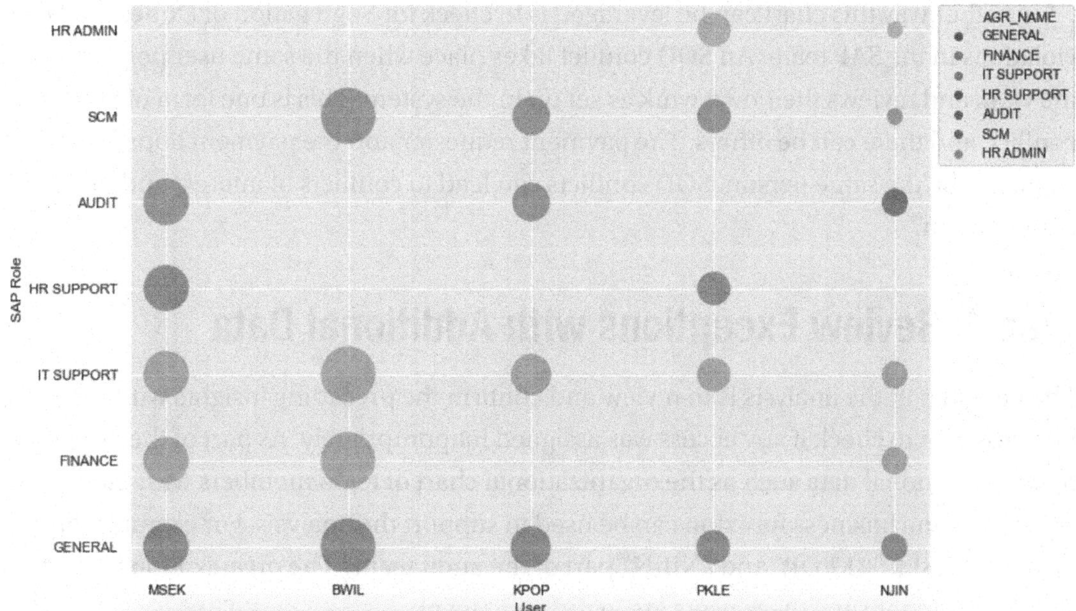


***Figure 22-1.*** *Scatter plot showing user role vs. username*

The following insights can be derived from the scatter plot illustrated in Figure 22-1 based on the sample data:

- User "BWIL" is the earliest user in the system. The bubbles are the biggest for this user.
- User "NJIN" possibly switched roles sometime later with the company. The size of the bubbles is smaller for HR ADMIN and SCM compared to the other roles possessed by user "NJIN."
- MSEK, KPOP, and NJIN appear to be part of the audit group.
- All users have the "GENERAL" and "IT SUPPORT" role.
- User "NJIN" appears to have almost all access in the system except for "HR SUPPORT."

These are just some insights that one plot provides. There are other insights that can be derived from this one plot alone. Although machine learning was not used for this analysis, it does use data science principles to gain insights into the data. Using a standard data analytics tool to get the same results can result in more data processing or analysis.

Another way this chart can be leveraged is to check for Segregation of Duties (SOD) violations in the SAP team. An SOD conflict takes place when the same user performs the work and reviews their own work as set up in the system. This is one form of SOD conflict, and there can be others. The payment requester and the payment approver shouldn't be the same person. SOD conflicts can lead to conflicts of interest and misuse of the system.

## Step 4: Review Exceptions with Additional Data

The next step in the analysis is to review and confirm the preceding insights with the business unit to check if any access was assigned inappropriately. As part of the auditor's review, additional data such as the organizational chart of team members with their corresponding business function can be used to support the analysis. For example: are all users "MSEK," "KPOP," and "NJIIN" part of the audit team? The organizational chart should help confirm if these users are in the audit team.

## Step 5: Reperform the Analysis

The analysis will need to be performed regularly to ensure that access changes are analyzed for access violations. One advantage of this type of simple analysis is its ability to capture users and roles of the system of much bigger sizes. Since most of the functions and libraries are built-in Python libraries and functions, the analysis is efficient. The frequency of the analysis depends on the size of the organization and the amount of access changes as well as the risks identified. If there has been a history of access violations, it can be run more frequently as an added control to existing access controls.

# Variation and Serving

For the plot, the size of the figure may need to be adjusted for a larger number of users or roles. This can be done by changing the environment option `figure.figsize`, as indicated in the provided code for the scatterplot. The size scale may need to be adjusted for a different dataset. Other dimensions of data can be used in the form of marker shapes. Where each marker shape can indicate a different dimension. For instance, the administrators of the system can be shown with a different marker shape using the `markers` option:

```
sns.scatterplot(...data = df, alpha = 0.5, size..)
```

The `alpha` option can be adjusted to make the points more or less transparent. For example, the preceding code makes the scatter plot 50% transparent. This can be helpful when having multiple overlapped points. The name of the role in the dataset is assumed to start with the business function in order for the analysis to work properly for easier identification. Not all SAP systems are configured this way. One way to overcome this is by compiling a new column that keeps track of the corresponding business function. The column containing the business function can be used to color the data points.

The same code can be leveraged for other ERPs and other configurations of SAP through relinking the fields. For instance, in another ERP, the role, level of access along with the start dates may be in separate tables. The tables will need to be joined in this case to make it look like AGR_USERS before the analysis can be re-performed. Another point mentioned earlier was that some organizations may employ UBAC. The analysis can be leveraged for UBAC as well.

Currently, the analysis assumes that all roles do not have an end date. In other words, this means no employee has left the organization. In reality, this is not the case and employees who leave the company will have an end date for the roles. This can also be seen when employees switch roles in the company. For a more inclusive analysis, the `date_convert()` function can be modified to take in the end date as another argument. If the end date is "9999-12-31," then the number of days of access would be the difference between the current date and the start date. If the end date is not "9999-12-31," then the number of days of access is the difference between the end date and the start date of the access.

Ideally, the analysis can be automated to be run on a regular basis and an image of the scatter plot can be published to a central place where an assigned auditor or group of auditors can go to review access. This analysis can provide additional support when auditors are working on an audit, or they can be reviewed on an ad hoc basis. The figure can be saved using the `plt.savefig()` function.

# CHAPTER 23

# Project Management

In this chapter, we will look at two ways to leverage ML when it comes to analyzing the success of the projects. A classifier is used to predict if a project will be successful. We will also look at a technique to determine feature importance. The techniques discussed in this chapter will help an auditor determine the projects that are likely to fail which can be great points for discussion with the business. The results of the feature importance analysis can be used to determine the variables that are most important when it comes to failure of high-risk projects.

The chapter is organized like a recipe – goal of the recipe, ingredients, instructions, and variation and serving. The accompanying code is available at the GitHub repository specified in Chapter 20.

## The Dish: Project Portfolio Analysis

Every organization has a portfolio of programs and projects. Projects and programs help organizations to work on something new with an end date to serve the organization's vision. According to the Project Management Institute's "Pulse of the Profession 2020" report, over 20% of projects fail in organizations with low project management maturity. An organization with a low maturity project management framework has ad hoc processes when it comes to managing its projects. In terms of value delivery, only about half of the projects meet goals/intent of the project, are within budget, time, and deal with scope creep. Scope creep occurs when the original requirements of the projects are revised throughout the course of the project. It would be beneficial to understand how the project metrics such as budget, time, met goals, and scope creep affect project failures.

Although internal audit is not directly responsible for the organization's project management, they are regularly engaged to perform audits that look at how effectively a project portfolio was managed. Auditors review the budget, time, scope, and other



© Maris Sekar 2022
M. Sekar, *Machine Learning for Auditors*, https://doi.org/10.1007/978-1-4842-8051-5_23

project management metrics of projects that pose a high risk to the organization. The feature importance analysis discussed below will help the auditor determine the most helpful metrics in predicting the success of a project. The insights from this analysis can be used as points for discussion during the risk assessment with the business units.

In this analysis, we will look at two ways of applying ML in project management. Using a Random Forest Classifier ML algorithm, we will predict if a project will be successful based on the budget, time, team size, and business function info. We will also look at which features are of most significant importance when it comes to predicting if a project will be successful using Random Forest's built-in variable importance capability.

# Ingredients

A sample dataset has been provided in the file Project_Portfolio.csv. Here are the columns along with their descriptions:

***Table 23-1.*** *Column listing of Project_Portfolio.csv*

| Column | Description |
|---|---|
| Project | Name of the project |
| Business unit | The business function the project belongs to. |
| Expected duration | The budgeted time for the project |
| Actual duration | The actual time taken to complete the project. |
| Budget amount | The amount budgeted for the project before the project commenced. |
| Actual amount | The actual cost of the project to the organization |
| Team size | The total size of the project team |
| Result | Indicates if the project was successful. |

# Instructions

There are four steps you'll want to follow. First is preparing the data. Then you'll want to do some exploratory data analysis to clarify your understanding of what you have to work with. Then apply a classification and review the results.

## Step 1: Data Preparation

Upon initial inspection of the columns, they were found to contain categorical values (for instance, Business unit, Result, Project, etc.). In order for the Random Forest algorithm to pick up the signals from the dataset, we will need to convert the categorical values to numerical values using One-Hot Encoding (OHE). Recall that OHE converts every category in a categorical field to a column and assigns a binary value of one or zero. In this analysis, the built-in pandas function `get_dummies()` is used to perform this task.

There are no repeating projects or projects with the same names. The uniqueness of the projects does not help with the classifier algorithm because it doesn't provide any further information other than that each of the lines is unique. So, the project column is first removed from the dataset.

It is observed that the "Result_Not Successful" and "Result_Successful" columns represent the target value we are trying to predict. Hence, these columns are dropped from the dataset.

## Step 2: Exploratory Data Analysis

As part of Exploratory Data Analysis (EDA), the summary statistics for all the columns are observed. Here are some interpretations of the summary:

- "Expected duration" ranges from three months to two years when we look at the min and max values for this field.
- "Actual duration" ranges from six months to four years when we look at the min and max values for this field.
- On average, there are ten team members in a project. This can be seen from the mean value of "Team Size" field.

## Step 3: Perform Random Forest Classification

A Random Forest classifier constructs a specified number of decision trees with random sets of features for each tree. Every decision tree's performance is calculated, and the information added by each of the features is noted using the training dataset. In this analysis, a 50%/50% split is specified for the train/test split. The combination of all the constructed trees (1000, in this case) is used to predict the test dataset. The performance is measured using the f1-score, which is 46% for the overall classifier. The classifier's ability to predict a successful project is 59% (see f1-score in the classification report for "Successful").

One of the decision trees used in the model is illustrated in Figure 23-1.

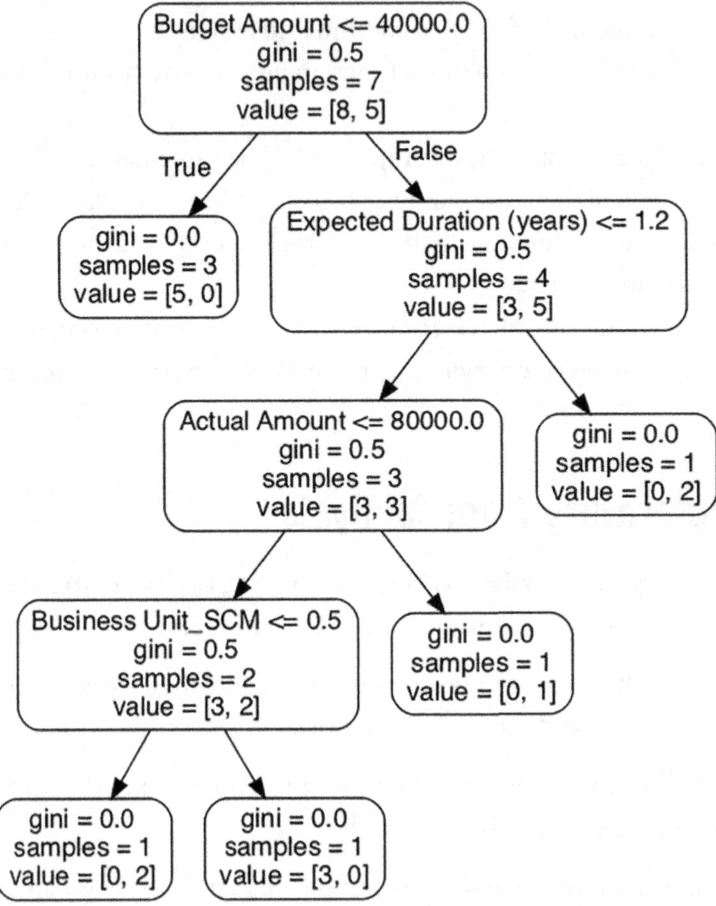


***Figure 23-1.*** *Decision tree used in the random forest classifier*

Figure 23-1 shows one decision tree out of 1000 decision trees used by the model. Each box represents a decision point or a node, and the arrows indicate the cases when it is true (left) and false (right). As an example, the very first node at the top checks if Budget Amount <= 40000. If the Budget Amount is less than $40,000, the Gini coefficient is zero. Gini coefficient is a statistical measure of the degree of variance in a dataset. A Gini coefficient of zero means there is no variance in the given dataset.

## Step 4: Review Feature Importance

The feature importance of the trained Random Forest model is retrieved via the `feature_importances_` parameter of the model. The feature importance is plotted using a bar plot, as seen in Figure 23-2.

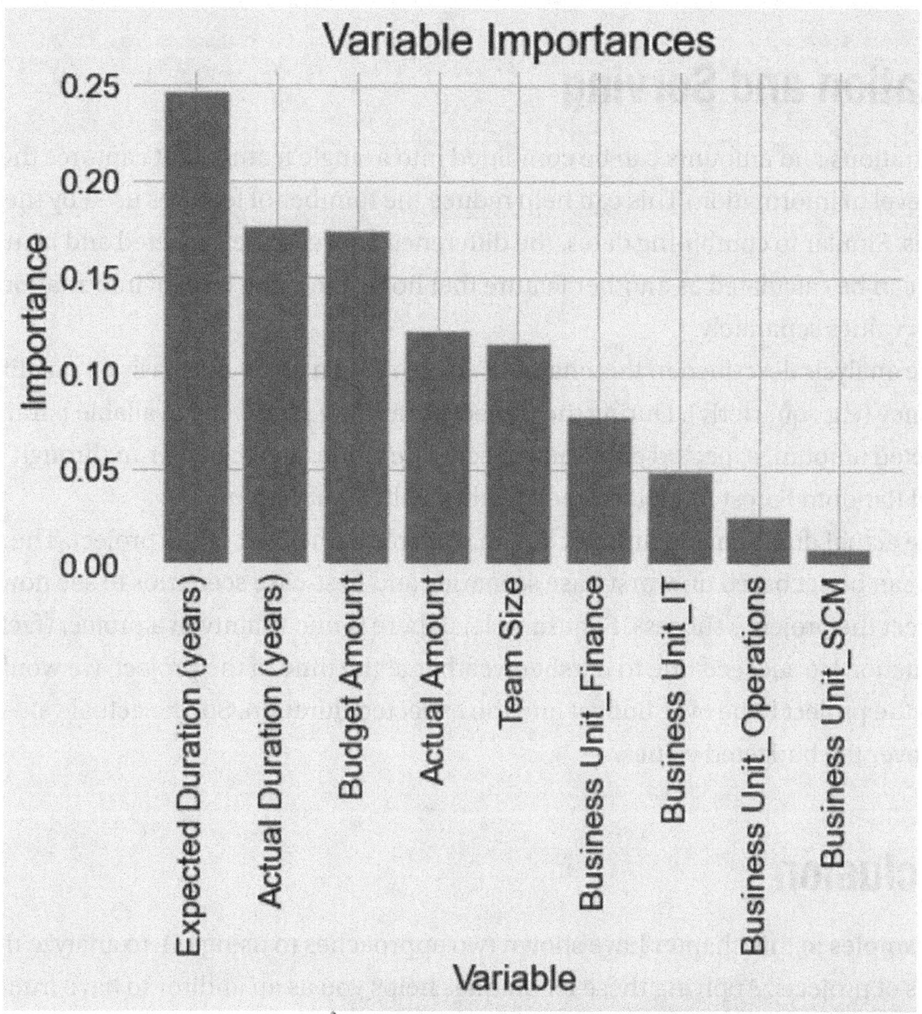


***Figure 23-2.** Variable Importances from the Random Forest Model*

It can be seen that Budget Amount has the highest importance at 25%. This means that when predicting if a project is successful or not, the "Budget Amount" variable captures 25% of the variance in information. The features "Budget Amount" (25%), "Expected Amount" ("18"), "Actual Duration" ("18"), "Actual Amount" ("12"), and "Team Size" ("12") contain 85% of the information.

## Variation and Serving

The durations and amounts can be combined into a single feature that captures the same level of information. This can help reduce the number of features used by the analysis. Similar to combining dates, the difference between the expected and actual values can be calculated as another feature that holds the same level of information as the two values separately.

The analysis described in this chapter can be used on an ad hoc basis or at a different frequency (e.g., quarterly). During the project's planning phase, the available parameters (budgeted amount, expected amount, expected duration, etc.) can be run through the trained Random Forest model to predict if they will be successful.

The actual duration and amount are unavailable at the start of the project. The actual values can be set based on worst-case scenarios and best-case scenarios to see how that will affect the project's success. For example, if there is uncertainty in a project (facility construction) to succeed due to harsher weather at the time of the project, we would expect the project to be over budget and the expected duration. So, the actual values can be set over the budgeted values.

## Conclusion

The examples in this chapter have shown two approaches to using ML to analyze the success of projects. Applying these techniques helps you as an auditor to have fruitful and sometimes difficult discussions with businesses around identifying projects that are high-risk and that merit special attention to guard against failure.

# CHAPTER 24

# Data Exploration

This chapter shows how we can apply exploratory data analysis (EDA) on raw data to gain insights from the data. A sample dataset is explored with some of the most commonly used methods. Techniques to check for missing values, frequency of occurrence, and the correlation between variables are shown in this analysis.

The chapter is organized like a recipe – goal of the recipe, ingredients, instructions, and variation and serving. The accompanying code is available at the GitHub repository specified in Chapter 20.

## The Dish: Understanding the Data Through Exploration

EDA is the analysis of data by summarizing its major components using data visualization and statistical methods. It is one of the first methods used when using data for ML or data science. In order to understand raw data, data can be visualized and inspected for missing values to determine correlations in the data. In this analysis, we will explore the publicly available aggregated COVID-19 case data provided by the Government of Alberta.

## Ingredients

The data used for the analysis is aggregated case data about COVID-19 cases reported to the Alberta Health Services for the period of March 6, 2020, to January 6, 2021.



© Maris Sekar 2022
M. Sekar, *Machine Learning for Auditors*, https://doi.org/10.1007/978-1-4842-8051-5_24

***Table 24-1.*** *Column data for covid19dataexport.csv*

| Column | Description |
| --- | --- |
| Unnamed column | The first column is unnamed and holds an auto-generated number. |
| Date reported | The date the case was reported. |
| Alberta Health Services Zone | The regional zone |
| Gender | Gender of the patient |
| Age group | Age range |
| Case status | Indicates if the patient recovered or if the case is still active as of the data extraction date. |
| Case type | Indicates if the case is confirmed. |

# Instructions

There are just two broad steps prior to exploratory data analysis. First you prepare the data. Then you perform the actual exploration.

## Step 1: Data Preparation

The first data preparation step is the renaming of columns. The first unnamed column is given a name so it can be easily identified. The other columns are mostly renamed to smaller or more sensible column names.

The other data preparation used in this analysis is label encoding, which is similar to one-hot encoding. One-hot encoding (OHE) is transforming data to prepare it for an algorithm and optimize the prediction. In label encoding, each categorical value in a column is converted into a numerical value. For instance, the "Gender" field has three categories: Male, Female, and Unknown. Here Male is assigned a zero, Female is designated as a one, and Unknown is set with two.

## Step 2: Exploratory Data Analysis

After the data is loaded into Jupyter Notebook, the shape of the data frame is retrieved. There are a total of 108,469 reported cases with seven characteristic variables.

The "Date Reported" date is next checked for the earliest and the latest reported date for the provided dataset. The earliest reported date is March 6, 2020, and the latest reported case was on January 6, 2021. This confirms the expected period of the dataset.

The data is checked for duplicates by using the built-in `duplicated()` function. A duplicate occurs when two or more rows have the same values. Duplicates can affect the performance of ML algorithms. There are no duplicates in the dataset.

Missing values are checked using a function named `count()`. The count function displays the total values found in all the columns. Null values can be shown using the `isnull()` function. In the provided example, there are no missing values. If missing values are found, they should be replaced by placeholders. This can be done using the `replace()` or `fillna()` functions. Missing values can negatively influence the algorithm, so removing them or filling them with expected values is better. Another method commonly used to replace empty numeric variables is replacing all null values with the average value calculated with the other available values.

The value_counts() function helps to understand the breakdown of the data by the various columns. The results are visualized to illustrate the counts clearly. Here is the summary of the columns after running the value_counts function:

- Status. The majority of the reported cases that have recovered are compared to active and dead cases. The dead cases are a small fraction compared to the Status.
- AHS Zone. Edmonton and Calgary Zones reported over 40,000 cases. North, Central, and South Zones reported much lower cases, each at around 6000–8000 cases.
- Age. Middle-aged people (30–39 and 20–29 years) have the highest reported cases. Children under one year have the lowest reported cases.
- Date reported. Spring and Summer months (April–August) have a lower number of cases. The reported cases are the highest in winter. December has the highest number of cases at about five times more than the summer months.
- There are as many male patients as female patients, at around 54,000 cases each.

A correlation matrix is constructed at the end to see the correlation between all the possible variable pairs provided in the dataset. It is best measured in variables that have a linear relationship with one another. A heatmap and a pairs plot are typically used to show the relationship between the variables. There are no significant correlations in the dataset. There is some negative correlation between Age and Status, but this is very small, as can be confirmed by the pairs plots.

# Variation and Serving

In this analysis, the method of imputation was described earlier. Imputation is the method of replacing null or empty values with alternative values. Discarding missing values may lead to bias or misrepresentation of the data. Suppose we measure the height of students in the classroom. There are seven students in the class, and their heights range from 150–170 cms. Three students are absent today, and their heights couldn't be measured. If only the seven students' heights were used, we would have only a 70% representation of the total possible data of the heights of the students.

A better way to make up for the missing students would be to set the height of the absent students to the average of the seven students in the class, which is 160 cms in this case. This is a better dataset to work with. The mode of the variable is another method that can be used to impute the values. The mode is the value with the most significant number of occurrences. If the largest number of height occurrences is 162 cms, this would be the mode of the height variable.

The script for data exploration can be standardized as a series of tests performed on any dataset being analyzed. The techniques covered in this chapter are some of the most commonly used methods. A more comprehensive list of data exploration tests can be developed and leveraged for understanding the data.

# Conclusion

You've seen in this chapter how to apply exploratory data analysis to gain insights from raw data. Such analysis is valuable in looking for and understanding the reasons behind missing values, frequency of occurrence, and any other anomalies that you might uncover during the exploration step.

CHAPTER 25

# Vendor Duplicate Payments

This chapter/recipe shows how the auditor can identify duplicate payments using the k-nearest neighbors (KNN) algorithm. After identification, the validation method of duplicate payments and interaction with the business will be explored.

The chapter is organized like a recipe – the goal of the recipe, ingredients, instructions, and variation and serving. The accompanying code is available at the GitHub repository specified in Chapter 20.

## The Dish: Vendor Duplicate Payments Analysis

Payments are made in error to the vendors in almost all organizations. One common problem is duplicate payments made to the vendor due to a flaw in the payments process. Consider an example where a vendor changes their address. A new record is created in the vendor master table to capture the new address. If there was a gap in the payments process that still paid the vendor in the old address, the vendor might call the organization after they realize they haven't received a payment. The organization might end up re-issuing another cheque to pay the vendor at the new address. If the original payment was not tracked and canceled on time, this could lead to duplicate payments.

In this analysis, we will use k-nearest neighbors with KD Tree indexing. The main idea behind k-nearest neighbors is to find n number of training samples closest in distance to a given point and predict the label of the new point. With KD Tree neighbors searches, the distance calculations are reduced by encoding the aggregate information of the sample in K dimensional trees.

This technique is to be used as a supplemental tool for internal auditors on top of their testing, controls, or processes in place to avoid duplicate payments.



© Maris Sekar 2022
M. Sekar, *Machine Learning for Auditors*, https://doi.org/10.1007/978-1-4842-8051-5_25

## Ingredients

For this analysis, 100,000 random samples with ten attributes were generated using a normal distribution. These ten attributes represent ten numerical fields from a payments table.

## Instructions

There are three steps in the analysis: preparing the data, executing the algorithm, and reviewing any exceptions. The following subsections describe each step briefly.

### Step 1: Data Preparation

The only data preparation used in this analysis is the addition of a new column. The new column keeps track of the Euclidean distance between the maximum of the normalized values of all columns and the normalized value of each column.

### Step 2: Perform K-NN Algorithm

The K-NN algorithm is trained using the 100,000 random samples with 11 (including the added column from Step 1) attributes with K=1 (one neighbor). KD Tree indexing is selected for training the model. After the model is fit to the data, we are ready to check if a given payment is similar to any of the current payments that the model has already seen.

A random payment sample is generated to check for its distance from any of the trained points. The distance and the index of the point with the minimum distance are displayed. In our analysis, the payment represented at Index 50753 has a minimum distance of 4.89.

### Step 3: Review Exceptions with Additional Data

The identified duplicates can be checked by retrieving the original purchase order, payment, and invoice details. The purchase order and invoice amounts must match the payments made to the vendor.

# Variation and Serving

The randomly generated payment transaction table can be substituted for the actual payment transactions. The payment transactions can be loaded into the Jupyter notebook environment using the `read_csv()` function. The attributes in the payment transactions will then need to be processed so that only the numerical variables are used for the analysis. If there is a categorical variable, it can be transformed into a numerical variable through label encoding. Please refer to the previous chapter on how to do this.

The duplicate payments can be identified in near real time using this algorithm. Every time a payment is made or every batch of 50 payments can trigger the K-NN model to be retrained with the new payments. Every new payment's distance can be calculated quickly to check if a duplicate payment was made. A threshold for the distance can be applied to control the false positive numbers. The higher the threshold, the lower the number of false positives. The f1-score can be used to determine the accuracy of classifying a duplicate payment correctly. The f1-score is a harmonic measure to calculate the accuracy based on precision and recall.

F1-score =2*(precision*recall/(precision + recall))

# Conclusion

Duplicate payments are of course not ideal for any business. Applying the k-nearest neighbors algorithm, as shown in this chapter, can help you to home in on possible duplicates, which you can then verify and resolve with the business.

# CHAPTER 26

# CAATs 2.0

This chapter illustrates the application of ML for Computer Assisted Auditing Techniques. The recipe will help analyze the transactions to gain additional hidden insights into the data and the business.

The chapter is organized like a recipe – goal of the recipe, ingredients, instructions, and variation and serving. The accompanying code is available at the GitHub repository specified in Chapter 20.

## The Dish: CAATs Analysis Using ML

Computer Assisted Auditing Techniques (CAATs) have been increasingly popular over the last decade. We introduce two main use cases of CAATs: Verification of accounting data and Check control effectiveness. In this analysis, we will explore a way to improve the current process of verification of accounting data. Accounting data such as General Ledgers (GL) and Trial Balances (TB) need to be balanced on a regular basis due to the volume of transactions and the possibility of fraud. GL is a ledger used for bookkeeping and contains transactions posted from journals from accounts payable, accounts receivable, fixed assets, purchasing, etc. The entries recorded in GL are at the individual transactional level. TB is a listing of all accounts along with their opening and closing, and credit and debit balances for a specified period. The TB entries are at the accounts level.

GLs and TBs are typically reconciled by auditors as a validation tool to ensure that company transactions are balanced throughout the course of the year. Since the entries are recorded at the transactional level (GL) and the account level (TB), reconciling them is a good check for fraud. If a transaction was added to the GL but the change was not reflected in the TB at the account level, the GL and TB will not reconcile.

Here is the general formula used to reconcile GL and TB:

$$\text{Difference} = (\text{GL\_Credit} - \text{GL\_Debit}) - (\text{TB\_Opening\_Balance} - \text{TB\_Closing\_Balance} + \text{TB\_Credit} - \text{TB\_Debit})$$



© Maris Sekar 2022
M. Sekar, *Machine Learning for Auditors*, https://doi.org/10.1007/978-1-4842-8051-5_26

Before the preceding formula can be applied, the GL will need to be transformed into account level summarized data. This can be done by summarizing the GL transactions at the account level and adding up the balances for credits and debits.

In the current process, the transaction details are checked only when there is a difference, that is, when the GL and TB do not reconcile. The GL descriptions, the text that is attached to each of the transactions describing the transaction, is often not used for any purposes.

In our analysis, we assume that we have the Difference (see earlier) and the GL description topics assigned through topic modeling. Topic modeling is a Natural Language Processing (NLP) technique that trains on unstructured data like text descriptions to assign topics through clustering keywords. An example of topic modeling is provided in the next chapter, Chapter 27.

Given the two features, our analysis is able to use K-Means clustering to segment the accounts based on the topics and their GL vs. TB differences. Another goal of the analysis is to identify anomalies present in the dataset.

# Ingredients

For this analysis, we create a random dataset with 200 observations and two features with the `make_blobs()` function. The `make_blobs()` function creates data points around a specified number of blobs (five in this case) and provides the corresponding cluster number of the data points. Five blobs were generated because they represent the five types of GL accounts: assets, liabilities, incomes, expense, and capital. The clusters (account types in this case) are important to know because it can be compared later to check if what we found through our clustering algorithm is accurate.

The two features represent topic number and differences. The topic number is assigned based on performing topic modeling on the GL descriptions, which will not be covered in the analysis. The details in the next chapter (Chapter 27) on topic modeling can be leveraged to run topic modeling on the GL descriptions.

# Instructions

There are three steps to CAATS analysis. As always, one begins with preparing the data.

## Step 1: Data Preparation

The first data preparation task is to scale the features. Standardization (or scaling) is especially important for clustering algorithms because it helps to reduce the dominating effect of variables with larger scales when defining clusters. This is the only major transformation in this analysis.

## Step 2: Exploratory Data Analysis

As part of the EDA process, the random dataset is visualized to see how the five clusters are distributed across the data points. The distribution of the clusters illustrates how the accounts are segmented. Each of the clusters represents a type of GL account.

## Step 3: K-Means Clustering

The k-means clustering algorithm is applied to the dataset to determine if there are any hidden relationships when the differences and descriptions are analyzed. A cluster size of three (K=3) is first used for analysis. k-means clustering assigns a cluster to the closest datapoint based on its distance from the center of the clusters. The center of the cluster is moved iteratively so that it always moves to the center of the datapoints assigned to the cluster.

The optimal number of clusters can be determined using the elbow method, which is three. The elbow method computes the Sum of Square Error (SSE) for cluster sizes from one to ten. The optimal cluster size is found at the point where the plot makes the steepest transition to the next size. In this case, three was found to be the optimal number of clusters since a size more than this (K=4) would provide only a small decrease in SSE.

When we created the data, we actually had five clusters - one cluster to represent one type of GL account type. But our analysis shows that there are only three distinct clusters, as shown in Figure 26-1.

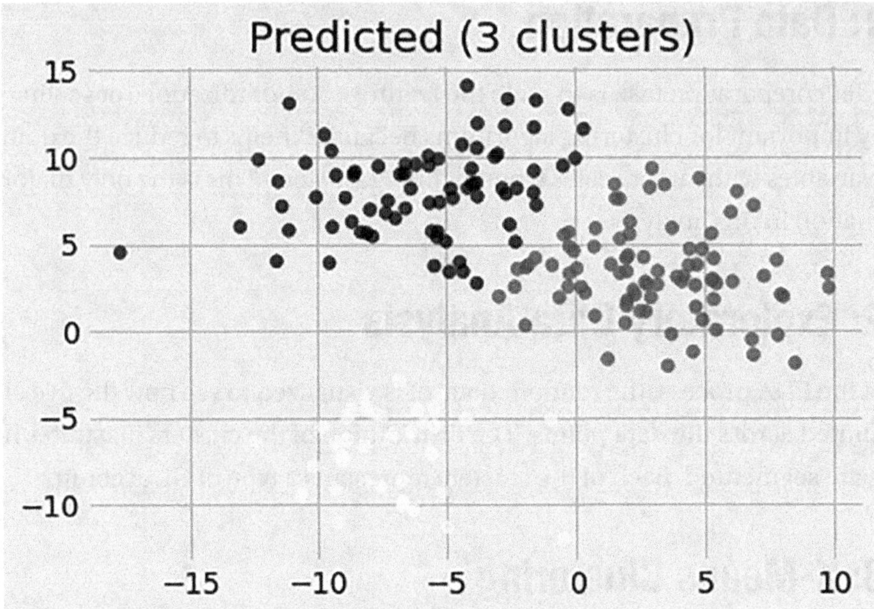


*Figure 26-1. Results of k-means clustering (Using K=3)*

One reason could be that the red and black clusters in the "Predicted (3 clusters)" plot can be further split into two other clusters which are not necessarily clear due to their proximity with each other.

Anomalies can be identified by their distance from the center of the cluster. In our analysis, one possible anomaly is the data point at (-15, 5). This data point, although assigned to a cluster, is far away from the rest of the points in the same cluster. The GL transactions associated with this data point can be explored further to determine if they are legitimate transactions by looking at additional data such as purchase order, invoice, and/or payment.

## Variation and Serving

When using topic modeling, the GL transactions data can be summarized and the GL descriptions of the associated transactions can be concatenated to associate it at the account level. See Figure 26-2 for a visual illustration of the summarization.

| Date | GL Description | Account Number | Amount | ... |
|---|---|---|---|---|
| 1/3/2021 | Item 1 | 10002 | 100.00 | ... |
| 2/1/2021 | Item 2 | 10002 | 125.00 | ... |
| 4/1/2021 | Item 3 | 10005 | 320.00 | .. |

| Account | Combined GL Description | Amount | ... |
|---|---|---|---|
| 10002 | Item 1, Item 2 | 225.00 | ... |
| 10005 | Item 3 | 320.00 | ... |

***Figure 26-2.*** *Summarization of GL transactions*

Other features such as posting date, transaction date, clearing date, currency, etc. can be used for segmentation. In order to prevent the selection of features with redundant information, Principal Component Analysis (PCA) can be used. PCA reduces the number of features and helps us avoid redundant features that do not add additional value.

One way to leverage dates in the current analyses is by calculating the difference in dates to get another feature that can be used in the analysis instead. For example, the k-means algorithm requires all features to be numerical. In order to

utilize the information contained in the dates field, a new field that shows the difference between the clearing date and the posting date can be calculated and used as a feature for the K-Means algorithm.

The final model for this analysis can be run on a regular basis (quarterly, semi-annual, or annual) to identify anomalies and understand the business better. A real time implementation of the analysis is likely not necessary for this application because CAATs are performed on a quarterly, semi-annual, or annual basis. However, it can be easily transformed into a continuously evolving model that is retrained on a daily or weekly basis depending on the volume of transactions.

# Conclusion

This chapter has shown how you can apply CAATs to help analyze transactions and gain insights that you can report back to the business. Such techniques can be used to verify accounting data and the effectiveness of the various accounting controls which an organization has implemented.

# CHAPTER 27

# Log Analysis

This chapter looks at the analysis of system and application logs generated by the organization. It shows how logs can be used to derive insights, gain a better understanding of the business, and identify high-risk areas of systems and applications. The knowledge of the high-risk regions can be used in internal audit's risk assessment process.

The chapter is organized like a recipe – the goal of the recipe, ingredients, instructions, and variation and serving. The accompanying code is available at the GitHub repository specified in Chapter 20.

## The Dish: NLP Log Analysis

Logs are created in almost all organizations in a variety of ways. They can be used to record activity on a server or application. The recorded activities are used as audit trails to determine what happened on an application at a given period of time. The Cybersecurity team uses logs as a way to determine if malicious actions took place. The engineering team can also use logs to detect and repair maintenance-related issues. In this analysis, we will look at how logs can be leveraged by the internal audit team to look at high-risk areas of an application using topic modeling. Topic modeling is a type of Natural Language Processing (NLP) algorithm that predicts the topics of a given set of text. In this analysis, topic modeling is used to predict the types of logs generated by a given log. For example, it would be helpful if all the lines in the log can be flagged if they were completed successfully or failed or some other event occurred. This analysis is valuable to IT auditors to narrow down the areas showing consistent challenges to the business. It could also be used as a way to confirm existing findings from another area of the audit.



© Maris Sekar 2022
M. Sekar, *Machine Learning for Auditors*, https://doi.org/10.1007/978-1-4842-8051-5_27

# Ingredients

Each log entry in the test data consists of four columns separated by the pipe operator ("|"). Here is a sample entry in the sample log:

**2021-01-01 08:20:00|MKSEK|"Starting process"|"Process A"**

Columns with more than one word are typically enclosed by ("). This tells the text parser within Python that the combination of words is part of a single column.

The headers for these columns are datetime, user, log, and process.

***Table 27-1.*** *Column Listing of test.log*

| Column | Description |
|---|---|
| datetime | Date and time of the log entry |
| user | Username of the user who generated the event. |
| log | The description of the event |
| process | The name of the process associated with the event. |

# Instructions

There are four steps to this analysis. First is data preparation. Next comes the exploratory analysis, which itself can be broken into three phases or parts. Then comes topic modeling, followed by possibly reperforming the analysis.

## Step 1: Data Preparation

The log data is loaded into a data frame and assigned with the corresponding headers using options within the `read_csv()` function.

The data preparation step for this analysis includes a filter on the logs for specific keywords we are expecting to find in the logs. In this case, we filter the log entries to show only the ones that contain "Fail," "Error," "Start," and "Complete."

Finally, all the letters in the log descriptions for all log entries are converted to lower case letters. This ensures that if there is an entry containing the word "Fail" and another entry containing "fail", they are still counted as the same keyword and not marked as two different words.

## Step 2: Exploratory Data Analysis

The exploratory data analysis is broken into three main parts. First, a word cloud is generated based on the descriptions of all the log entries. A word cloud is a graphical representation of recurring words in a given text. Although word clouds have been heavily used over the last decade, one underused feature when using word cloud is "stop words". "Stop words" are a list of words you tell the word cloud generator to ignore when plotting them to the screen. Some stop words are "the," "is," "are," etc. This is essential in order to weed through millions of lines of log entries that can potentially generate hundreds of thousands of words. Many NLP libraries now come equipped with pre-selected stop words that can be used to filter out commonly used English words. In many cases, additional stop words may need to be specified based on the nature of the logs being reviewed and the message being conveyed. For example, in our sample log, the words "process" and "run" are overused and do not add any value by showing up in the word cloud. So, these words have been added as stop words.

The word "starting" has been seen as being used the most number of times in the sample, followed by "completed," "successfully," "errors," "encountered," and "failed." This tells us that there are as many errors seen in the logs as there is the number of completions.

The next part of the analysis focuses on n-grams. N-grams work similarly to word clouds in that they keep track of the frequency of the words. The frequency is often represented as histograms. In our analysis, we performed a unigram (1 word) and a bigram (2 words) analysis. The unigram results verify what we found in the previous analysis with the word clouds, but it now tells us the exact frequency of the words. The bigram analysis suggests that "completed successfully" is the most commonly recurring word combination within the log, followed by "errors encountered," and "encountered failed." We can see that there is a theme emerging with successful and failed process entries.

The final part of the analysis shows a scatter plot (shown in Figure 27-1) with the Process associated with the log entry on the Y-axis and the corresponding time it was generated on the X-axis. The log entries with the "errors" keyword are filtered and plotted in the scatter plot.

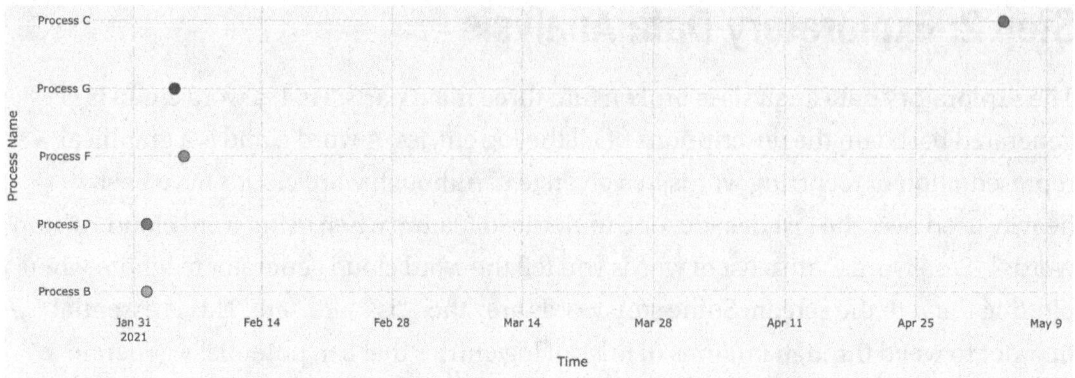


***Figure 27-1.** Scatter plot showing processes along with their generated time*

Figure 27-1 illustrates that all processes have failed except for process A and process E. Most of the failed processes occurred at the beginning of February. This finding may be a good talking point with the auditee to further confirm the understanding of the business. If the auditee has never seen this before, the same finding can become a value-added insight the business can leverage to better address gaps in their process. The auditor can strengthen their role as a trusted advisor by bringing such simple but powerful insights to the table.

## Step 3: Perform Topic Modeling

In the topic modeling step, the actual NLP machine learning model is generated by building a Latent Dirichlet Allocation (LDA) model from the sample data. The algorithm tries to come up with distinct topics along with the combination of words seen in each of the topics with their frequency. It is interesting to note that when we select two topics, the first topic consists of the successful process events and the second topic comprises the failed process runs. The sample dataset shows how the LDA model has generalized the successful and failed runs from the log entry descriptions. This trained model can now be used on a larger dataset to assign successful and failed topics.

## Step 4: Reperform the Analysis

The LDA model will need to be retrained whenever the structure of the logs changes or the mechanism used to record logs undergoes an update. Log generation activities do not generally change, so the model may not need regular retraining. The results will still need to be monitored to ensure that the model is working as expected.

# Variation and Serving

Knowing what keywords to filter at the beginning of the analysis is a trial-and-error exercise. It involves experimentation to find the entries that can be ignored and those that can be used to solve the business problem at hand.

For this analysis and the sample data, the number of topics was specified to be two and this resulted in an accurate topic assignment. For other datasets, the number of topics would change. Exploration of multiple numbers of topics and their corresponding performance with the dataset will assist in determining a suitable number for the final model.

In terms of deployment, the model can be retrained on an ad hoc basis. The retraining can occur when there is a performance degradation with the original model or if the structure of the data changes. Once trained, the model can be deployed to production and this cycle can continue.

For high-risk scenarios, the preceding retraining can be automated and a more real time or continuously streaming application can be implemented. The application can be set up to leverage the LDA model to assign topics as new log entries are generated. The topics with a failed run may be escalated to the internal audit team for further review and follow-up.

# Conclusion

You've seen in this chapter an example on how to look at system and application logs. Analyzing such logs helps you to gain a better understanding of the business and sometimes can help you to identify high-risk aspects of systems and applications that require a closer look. Your knowledge of these high-risk aspects is also useful in your risk assessment process.

# CHAPTER 28

# Concluding Remarks

This book started with the emphasis on internal audit's role in adding and protecting business value to an organization. The Three Lines of Defense model highlights the case for internal audit to align its work to the prioritized interests of the organization. AI/ML solutions provide exciting opportunities to support internal audit's commitment to becoming trusted advisors by providing value-added insights.

AI and ML solutions need to overcome unique data challenges faced by internal audit. The challenges can be in the form of people, processes, and technology. Some of the technologies currently employed by internal audit from a data analytics point of view include CAATs, Process Mining, and Continuous Auditing. Data Analytics is also leveraged by auditors to perform descriptive and diagnostic analyses. The audit staff environment needs to be taken into consideration for incorporating data analytics. In particular, senior leaders play a crucial role in ensuring the appropriate data analytics resources are in place to support data analytics projects.

Next, a fundamental knowledge of AI, ML, and data science was discussed. A functional understanding of the domain is essential for the effectiveness of AI solutions. To perform predictive modeling, internal auditors need to leverage both data science expertise and domain knowledge. Auditors need to take on the position of "Trust, but Verify" to ensure that all findings are verified by the business. Data Lakes, Cloud, and OT technologies can be incorporated to design AI/ML solutions. These technologies improve the accessibility, scalability, and flexibility of AI systems.

Storytelling is a key distinction between data analytics and data science solutions. Effective stories set the context and use data visualizations to convey key messages clearly and concisely. The audit reporting process can benefit vastly by incorporating storytelling. Specifically, storytelling can be used to convey observations to a non-technical audience and to inspire the decisionmakers to take corrective action. Visualizations can be used to support observations and to present data in terms of interactive graphics.



© Maris Sekar 2022
M. Sekar, *Machine Learning for Auditors*, https://doi.org/10.1007/978-1-4842-8051-5_28

Finally, practical use cases, organized as cooking recipes, were shared at the end of the book. These recipes provide a practical step-by-step approach to solving common internal audit applications using AI/ML techniques. The recipes cover methods to support fraud and anomaly detection, access management, project management, understanding data, analyzing vendor payments, CAATs, and investigating logs. The results from these analyses need to be verified and confirmed by the business.

AI/ML solutions can be leveraged by internal auditors to provide business value to the organization. When data analysis techniques were first introduced in audits, they transformed the internal audit team to use data and visualizations to support observations. AI, ML, and data science will further metamorphose the internal audit function to become trusted advisors of the organization.

# Index

## A



## B



## C





© Maris Sekar 2022
M. Sekar, *Machine Learning for Auditors*, https://doi.org/10.1007/978-1-4842-8051-5

## D

## E



## F



## G



## H



## I, J

## K



## L



## M



## N



## O



## P

Printed in the United States
by Baker & Taylor Publisher Services